SERIES ON KNOTS AND EVERYTHING

Editor-in-charge: Louis H. Kauffman *(Univ. of Illinois, Chicago)*

The Series on Knots and Everything: is a book series polarized around the theory of knots. Volume 1 in the series is Louis H Kauffman's Knots and Physics.

One purpose of this series is to continue the exploration of many of the themes indicated in Volume 1. These themes reach out beyond knot theory into physics, mathematics, logic, linguistics, philosophy, biology and practical experience. All of these outreaches have relations with knot theory when knot theory is regarded as a pivot or meeting place for apparently separate ideas. Knots act as such a pivotal place. We do not fully understand why this is so. The series represents stages in the exploration of this nexus.

Details of the titles in this series to date give a picture of the enterprise.

Published:

K⊗E Series on Knots and Everything — Vol. 26

FUNCTORIAL KNOT THEORY

Categories of Tangles, Coherence, Categorical Deformations, and Topological Invariants

David N. Yetter

Department of Mathematics
Kansas State University

World Scientific
Singapore • New Jersey • London • Hong Kong

Published by

World Scientific Publishing Co. Pte. Ltd.

5 Toh Tuck Link, Singapore 596224

USA office: 27 Warren Street, Suite 401-402, Hackensack, NJ 07601

UK office: 57 Shelton Street, Covent Garden, London WC2H 9HE

British Library Cataloguing-in-Publication Data
A catalogue record for this book is available from the British Library.

Series on Knots and Everything — Vol. 26
FUNCTORIAL KNOT THEORY
Categories of Tangles, Coherence, Categorical Deformations, and Topological Invariants

Copyright © 2001 by World Scientific Publishing Co. Pte. Ltd.

ISBN-13 978-981-02-4443-9
ISBN-10 981-02-4443-6

To my wife, Georgette

Acknowledgments

No monograph would be complete without the acknowledgment of those who contributed to its production in ways short of authorship, but whose contribrutions are not discernable from the citations.

First, and foremost, I wish to thank my wife, Georgette, who is not only a constant source of encouragement and emotional support, but also the best mathematical copy-editor I know. Second, I wish to thank those mathematicians who have made useful contributions to this work, either as collaborators on previously published results found herein or as critics of early versions of the manuscript: Louis Crane, Peter Freyd, Murray Gerstenhaber, and Saunders Mac Lane. Finally, I gratefully acknowledge the National Science Foundation for providing financial support (grant # DMS-9971510) during the period when this monograph was being prepared, and Kansas State University and the University of Pennsylvania for providing the venues and computer equipment used in its completion.

David N. Yetter
Department of Mathematics
Kansas State University
Manhattan, KS 66503 U.S.A.

Contents

II   Deformations

List of Figures

5

Chapter 1

Introduction

One of the most remarkable developments in the recent history of mathematics has been the discovery of an intimate connection between the central objects of study in low-dimensional geometric topology—classical knots and links and low-dimensional manifolds themselves—and what had heretofore been somewhat exotic algebraic objects—Hopf algebras, monoidal categories, and even more abstract-seeming structures. A correspondence between geometric and algebraic structures has, of course, been central to the development of mathematics, at least since Descartes provided the world with the coördinate plane.

The connection disclosed by the rise of "quantum topology" is, however, of a different character from that classically known. The classical connection is mediated by algebras of (possibly quite generalized) functions, so that the correspondence between geometric and algebraic objects is contravariant, as, for example, the correspondence between manifolds and algebras of smooth functions, or between affine schemes and commutative rings. In the connections between topology and algebra which have come to light since the discovery of the Jones polynomial, the topological objects (usually parts or relative versions of the primary objects of interest) are themselves the elements of an algebraic object. Topological information is then wrung from this algebraic object by representing it in other algebraic objects of the same type.

Another feature of these recent developments is the difference between the role categories and functors have usually played since their discovery and the role they now play in quantum topology. Rather than serving a foundational role, as a clean way of encoding "natural" constructions of one kind of mathematical object from another, categories in quantum topology stand as algebraic objects in their own right. This difference has not always been generally understood, even by quite brilliant mathematicians working in related areas, as the following personal anecdote involving the late Moshé Flato illustrates. One evening at a Joint Summer Research Conference in the early 1990's Nicholai Reshetikhin and I button-holed Flato, and explained at length Shum's coherence theorem and the role of categories in "quantum knot invariants". Flato was persistently dismissive of categories as a "mere language". I retired for the evening, leaving Reshetikhin and Flato to the discussion. At the next morning's session, Flato tapped me on the shoulder, and, giving a thumbs-up sign, whispered, "Hey! Viva les categories! These new ones, the braided monoidal ones."

It is the purpose of this book to lay out clearly and in one place much of the scattered lore concerning the categories most intimately related with classical knot theory, and to relate these categories both to knot polynomials, which were the original motivation for their study, and to the theory of Vassiliev invariants. No claim is made that this treatment is exhaustive of the current state of knowledge, but it is the author's hope that it will prove useful to students and established researchers alike. One area specifically not touched in this work (though some of the requisite definitions are mentioned as examples) is the connection between the theory of monoidal categories and the known algebraic constructions of topological quantum field theories. We have also steered clear of any areas in which the universal constructions charateristic of category theory in its foundational role are needed, as for example limits or colimits of diagrams. By doing this, we emphasize the algebraic nature of the subject at hand.

Part I lays out the fundamentals of "functorial knot theory", recalling the necessary facts and theorems from both category theory and knot theory, and even providing proofs of some "folk theorems" which are universally assumed. Part II shows that Vassiliev theory, at least in its combinatorial guise, falls within the scope of functorial knot theory, and thus understood can be viewed as a species of algebraic deformation theory. Part I is intended to be fairly self-contained, with only standard topics in first year graduate courses as prerequisites. Part II assumes some familiarity with algebraic deformation theory (in particular, Gerstenhaber [23, 24] and Gerstenhaber and Schack [25]) and homological algebra (see, for example, Weibel [57]).

Part I

Knots and Categories

Chapter 2

Basic Concepts

In this chapter we introduce basic concepts from low-dimensional topology and category theory which will be required in this study. We will begin with concepts from classical knot theory, and then turn to categorical structures. Whenever possible, we will illustrate categorical notions with both of classically known "categories-as-foundations" examples, and with more recent "categories-as-algebra" examples, these latter being chosen to emphasize the close connection between the categorical concept and low-dimensional topology.

Throughout this study, unless otherwise specified, terms like "manifold", "map", "embedding" and "homotopy" will refer to the piecewise linear (PL) version of the concept. Due to various classical smoothing and triangulation theorems, it would generally be a matter of indifference if the smooth versions were being used. Although there are some concepts, such as framed links, which are more natural in the smooth setting, we prefer the PL setting to avoid some niceties involving restrictions on germs near boundaries which are needed to develop the theory of smooth tangles. We will address these in Chapter 8. In the earlier chapters we will attempt to point out the adjustments which would be needed in the smooth setting, either in asides or in footnotes.

Throughout this work the unit interval $[0, 1] \subset \mathbb{R}$ is denoted $\mathbb{I}$.

2.1   Knots, Links and Tangles

Knots and links, that is to say, compact 1-submanifolds of $\mathbb{R}^3$ or $\mathbf{S}^3$, play a remarkably important role in the theory of smooth or piecewise linear 3- and 4-manifolds, and in a variety of other parts of mathematics and the sciences.

When equipped with a framing (or in the presence of orientations, a smooth field of normal vectors), they provide the data for the attaching of 2-handles to $\mathbf{B}^4$. Theorems of Kirby [35] show that every compact oriented 3-manifold arises as the boundary of a 4-dimensional handle-body with only 0- and 2-handles, and provides a calculus of "moves" to relate any two presentations of the same (diffeomorphic) 3-manifold(s). Similarly, the 2-handle structure turns out to be central to the properties of smooth 4-manifolds.

Many properties of singularities of complex plane curves are intimately related to the "link" of the singularity, that is, the intersection of the curve with the bounding $\mathbf{S}^3$ of a sufficiently small ball about the singularity. Finite families of closed trajectories of 3-dimensional dynamical systems can form links of arbitrary complexity.

Bacterial DNA forms a closed loop, and is thus reasonably modeled by a knot. Certain enzyme actions lead to very complex knots. More remarkable still, knots and links arise naturally from considerations in the quantization of general relativity.

For all of these reasons, the study of knots and links is of great interest, and it behooves us to consider precise definitions:

Definition 2.1 *A* (classical) knot *is an embedding of* $\mathbf{S}^1$ *into* $\mathbf{S}^3$ *(or* $\mathbb{R}^3$*).*

A (classical) link *is an embedding of* $\coprod_{i=1}^{n} \mathbf{S}^1$ *into* $\mathbf{S}^3$ *(or* $\mathbb{R}^3$*), for some* $n \in \mathbb{N}$*. (Note: we include* 0*, so that there is an "empty link").*

In all of the applications noted above, and whenever knots and links are studied topologically, the important thing is not the embedding itself, but its class under a suitable notion of equivalence defined in terms of geometric deformations. The naive notions of geometric deformation,

homotopy, or even isotopy (that is, homotopy through embeddings) turn
out to be unsuitable. Therefore we make

Definition 2.2 *Two knots or links* K_1, K_2 *are* ambient isotopic *or sim-
ply* equivalent *if there is an isotopy* $H : \mathbf{S}^3 \times \mathbb{I} \to \mathbf{S}^3$ *(or similarly for*
$\mathbb{R}^3$ *instead of* $\mathbf{S}^3$*) which carries one to the other.*
    More precisely, H *is a PL map, satisfying* $H(-, 0) = Id_{\mathbf{S}^3}$*;* $H(-, t)$
is a PL-homeomorphism for each t*; and*

$$H(K_1(x), 1) = K_2(x)$$

(using K_i *to denote the mapping, with implied domain.)*

    In this study, it is important to consider also a "relative" or local
version of knots and links confined to a rectangular solid:

Definition 2.3 *A* tangle *is an embedding* $T : X \to \mathbb{I}^3$ *of a 1-manifold
with boundary into the rectangular solid* $\mathbb{I}^3$ *satisfying*

$$T(\partial X) = T(X) \cap \partial \mathbb{I}^3 = T(X) \cap (\mathbb{I}^2 \times \{0, 1\}).$$

    The relevant notion of equivalence for tangles is then given by

Definition 2.4 *Two tangles* $T_1 : X_1 \to \mathbb{I}^3$ *and* $T_2 : X_2 \to \mathbb{I}^3$ *are*
equivalent *or* isotopic rel boundary *if there exist a PL homeomorphism*
$\Phi : X_1 \to X_2$ *and a map* $H : \mathbb{I}^3 \times \mathbb{I} \to \mathbb{I}^3$ *satisfying*

1. $H|_{\partial \mathbb{I}^3 \times \mathbb{I}} = p_{\partial \mathbb{I}^3}$

2. $H(-, t)$ *is a PL homeomorphism for all* t

3. $H(-, 0) = Id_{\mathbb{I}^3}$

4. $H(T_1, 1) = T_2(\Phi) : X_1 \to \mathbb{I}^3$

    The following lemma about ambient isotopies in $\mathbb{I}^3$ will be useful in
what follows:

Lemma 2.5 *Given an isotopy H of a closed set $F = [\epsilon, 1 - \epsilon]^3 \subset \mathbb{I}^3$, there is an isotopy $\tilde{H}$ of $\mathbb{I}^3$ to itself whose restriction to F is H, and whose restriction to $\partial \mathbb{I}^3$ is the trivial isotopy $p_{\partial \mathbb{I}^3} : \partial \mathbb{I}^3 \times \mathbb{I} \to \partial \mathbb{I}^3$.*

proof: Consider triangulations of $F \times \mathbb{I}$ and F on which the H is given by linear maps of the simplexes. Now, choose triangulations of $\partial \mathbb{I}^3 \times \mathbb{I}$ and $\partial \mathbb{I}^3$ subordinate to which the projection is given by linear maps of simplexes. Subdivide these triangulations so that the triangulation of ∂F and the triangulation of $\partial \mathbb{I}^3$ are isomorphic by the map given by radial projection from the center of $\mathbb{I}^3$.

Now, $\overline{\mathbb{I}^3 \setminus F}$ is PL homeomorphic to $[\partial F] \times \mathbb{I}$. Choose a PL homeomorphism $\phi_1 \times \phi_2 = \phi : \overline{\mathbb{I}^3 \setminus F} \to [\partial F] \times \mathbb{I}$ with the property that $\phi_2(\partial F) = 1$ and $\phi_2(\partial \mathbb{I}^3) = 0$. Then there is a piecewise smooth isotopy $S : \overline{\mathbb{I}^3 \setminus F} \times \mathbb{I} \to \overline{\mathbb{I}^3 \setminus F}$ given by $S(x, t) = \phi^{-1}(H(\phi_1(x), \phi_2(x) \cdot t), \phi_2(x))$ whose restrictions to ∂F and $\partial \mathbb{I}^3$ are linear. Now, let Σ be a PL approximation to S agreeing with S on ∂F and $\partial \mathbb{I}^3$. The desired isotopy is then given by

$$\tilde{H}(x, t) = \begin{cases} H(x, t) & \text{if } x \in F \\ \Sigma(x, t) & \text{if } x \in \mathbb{I}^3 \setminus F \end{cases}.$$

$\square$

There are two particularly important auxiliary structures with which knots, links and tangles may be equipped: orientations and framings. The first may be defined either homologically or combinatorially in the PL setting.[1] We prefer the combinatorial approach:

Definition 2.6 *A knot, link or tangle is* oriented *if every edge is equipped with a choice of one of its vertices as "first", in such a way that no vertex is chosen as "first" for both edges with which it is incident. We encode this choice diagrammatically by equipping each edge with an arrow pointing from the first vertex to the other (last) vertex.*

---

[1]Of course in the smooth setting, we could also define orientations in terms of orientation on the tangent bundle.

Observe that it suffices to equip one arrow in each connected component of a knot, link, or tangle with an arrow to specify completely an orientation on it.

The second notion, that of framing, exists most naturally in the smooth setting as a choice of a framing for the normal bundle of the (smooth) knot, link, or tangle. We may, however, easily translate it into the PL setting as follows: in the presence of the standard orientation on the ambient $\mathbb{R}^3$, $\mathbf{S}^3$ or $\mathbb{I}^3$, and an orientation on the knot, link or tangle, the specification of a framing on the normal bundle can be reduced to the specification of a field of normal vectors, since a second normal vector may be obtained as the cross-product of the unit tangent vector with the given normal vector. Using the exponential map of the standard metric, we can replace this normal vector field with a thin ribbon, one edge of which is the knot, link, or tangle. We can then take this "ribbon" version of framed links and translate them into the PL setting:

Definition 2.7 *A framing of a (PL) knot, link, or tangle is an extension of the embedding $T : M^1 \to X^3$ (for $X^3 = S^3$, $\mathbb{R}^3$ or $\mathbb{I}^3$) defining the knot, link, or tangle to an embedding $T_f : M^1 \times \mathbb{I} \to X^3$ such that $T_f(x,0) = T(x)$, and (in the case of tangles) if $x \in \partial X^1$, then $T_f(x,t) \in \mathbb{I}^2 \times \{0,1\}$ for all $t \in \mathbb{I}$.*

In Chapter 8 we will consider the smooth approach in more detail.

We can also encode a framing by attaching an integer to each component of the knot, link, or tangle. In the case of knots and links, this integer is simply the linking number of the two boundaries of the ribbon (with the orientation on the opposite boundary reversed).

In the case of tangles, an encoding of framings by integers can be given, but either it will be non-canonical and involve a choice of which framing is the 0-framing for each interval component, or it will involve further restrictions on the intersections of the tangle with $\partial \mathbb{I}^3$.

In cases where we consider the tangles to be oriented or framed, we require that the ambient isotopy in the definition of equivalence respect the orientation or framing in the obvious sense.

In all cases, of knots, links, or tangles, with or without orientations or framings, the abuse of language which ignores the distinction between a thing and its equivalence class is commonplace. For example, "the unknot" refers to the equivalence class of a planar circle.

Although the fundamental topological notion of equivalence is that of ambient isotopy, or ambient isotopy rel boundary, it is convenient in practice to replace this notion with a more combinatorial notion. The relevant notion was given in the classic treatise on knot theory, *Knottentheorie*, by K. Reidemeister [44]:

Definition 2.8 *Two PL knots, links, or tangles are isotopic by moves if they can be related by a sequence of moves of the following form:*

> *Let Δ be a closed triangle (in some triangulation in the PL structure on $\mathbb{R}^3$, $\mathbf{S}^3$, or $\mathbb{I}^3$ as relevant) such that the intersection of the knot, link or tangle, T, is exactly one or two of the closed edges of Δ. Replace $\Delta \cup T$ with the closure of the edges of Δ not contained in T.*

We then have

Proposition 2.9 *Two knots (resp. links, tangles) T_1 and T_2 are equivalent if and only if they are isotopic by moves.*

In the case of knots and links, the proof is given in Reidemeister [44]. For tangles, Reidemeister's proof together with Lemma 2.5 give the desired result.

One important fact about knots, links and tangles is that they can be completely characterized up to equivalence by certain planar drawings, called "diagrams". A sequence of propositions and definitions make this precise:

Proposition 2.10 *Almost every (orthogonal) projection of a knot or link K onto a plane is "at-most-two-to-one", in the sense that the inverse image of any point of the plane contains zero, one or two points of K, with only (isolated) transverse double points. We call such a projection a* regular projection.

proof: The PL proof may be found in detail in [44]. We sketch it here. Observe that the (orthogonal) projections in $\mathbb{R}^3$ are parameterized by $\mathbf{S}^2$. "Almost every" then indicates all except a set of measure zero in $\mathbf{S}^2$, in particular, all projections except a family parameterized by a curve (perhaps with isolated points) in $\mathbf{S}^2$.

One must avoid the directions of the edges (a finite set of points) so that many-to-one image points do not arise by the projection of an edge to a point. For each pair of edges, the directions of secant lines joining a point of one edge to a point of the other form two (topological) disks or arcs on $\mathbf{S}^2$. In the case where they form arcs, we must avoid these arcs to ensure transversality of double points, and we must likewise avoid directions of secant lines from any vertex to any point for the same reason (a finite set of arcs and points). For each pair of edges, one must avoid directions of secant lines from a point on one edge to a point on the other which also hit other points, to avoid image points with multiplicity greater than two. The secant lines themselves fill a closed region of $\mathbb{R}^3$ in such a way that every point of the region, except those on the two edges, lies on exactly one secant line. We must thus avoid a curve of directions described by the intersection of the other edges of K with the region. $\square$

In the case of tangles, an analogous result holds, though here we wish to consider only projections onto the "back wall" of the cube $\mathbb{I}^3$. Therefore we consider non-orthogonal projections onto the plane of the "back wall" followed by linear scaling into a standard square.

Of course, information is lost in the process of projection: one no longer knows the height of the points above or below the plane of projection. Since we are concerned with knots, links and tangles only up to equivalence, most of the lost information is irrelevant: there are ambient isotopies (or isotopies by moves) which preserve the projection, but change the height of the points. What cannot be changed by an ambient isotopy that preserves the projection is which of the preimage points of a double point lies above the other.

In fact, it is the case that this information about the preimages of each double point is enough to recover the knot or link up to equivalence.

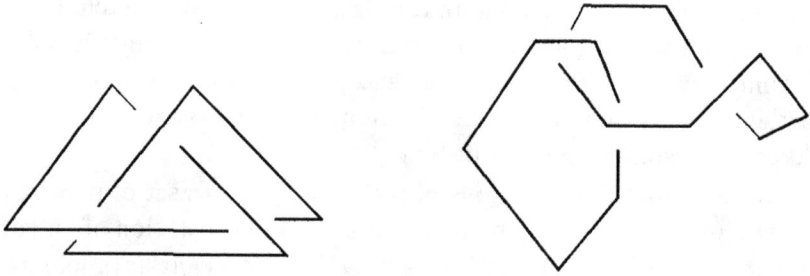

Figure 2.1: Examples of Knot Diagrams

By convention, the information is given by a *knot (or link) diagram*: a drawing of the projection in which the arc containing the lower of the two preimages is broken on either side of the double point, as, for example, in Figure 2.1. As is standard practice, we refer to these as knot diagrams, or simply diagrams, even in the case of links, and refer to the double points with the lower preimage indicated by the broken arc as *crossings*.

We then have

Theorem 2.11 *A knot or link is determined up to equivalence by any of its diagrams.*

The double points of a link diagram are called *crossings*. In the case where the link is oriented, we can distinguish two different types of crossings:

Definition 2.12 *Crossings in an oriented link diagram are* positive *or* negative *if the over- and under-crossing arcs are oriented as in Figure 2.2.*

Mnemonically, a crossing is positive if the right-hand rule curling from the out-bound over-crossing arc to the out-bound under-crossing arc gives a vector pointing up out of the plane of projection.

This then raises the question of when two diagrams determine the same equivalence class of knot of links. The answer is given by the classical theorem of Reidemeister [44]:

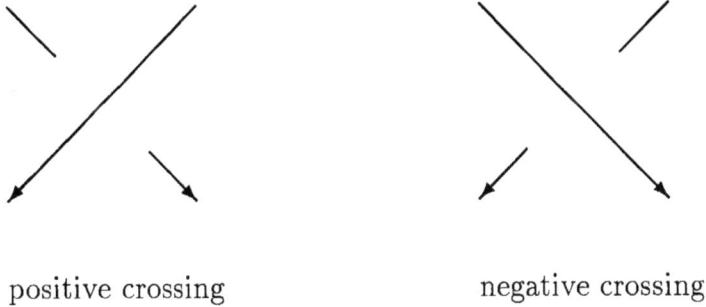

positive crossing   negative crossing

Figure 2.2: Crossing Signs

Theorem 2.13 *Two knot diagrams determine equivalent links if and only if they are related by a sequence of moves of the forms given in Figure 2.3.*

Before giving the proof of Theorem 2.13 we should comment on the fact that our set of moves is the original, larger set of combinatorial moves given in [44] rather than the smaller set, $\Omega.1$, $\Omega.2$ and $\Omega.3$, which is usually given under the name "Reidemeister moves" (cf. for example Burde and Zieschang [11]). The moves $\Delta.\pi.1$ and $\Delta.\pi.2$ are usually collected together in the phrase "isotopies of the plane of projection". Their inclusion, however, is both a convenience in the proof and, once the categorical structure of tangles is considered, a necessity for this study.

proof of Theorem 2.13:

The key to the proof is Reidemeister's other result: that ambient isotopy is equivalent to isotopy by moves. Consider a move across a triangle: if the projection of the triangle is an arc, the projection is unchanged by the move; otherwise, the projection of the triangle is itself a triangle.

To see that equivalence of diagrams under the diagrammatic moves implies isotopy by moves of the links is quite easy: each diagrammatic move becomes an isotopy-by-moves of the following form—use moves across triangles perpendicular to the plane of projection to adjust heights until the diagrammatic move can be realized as a single

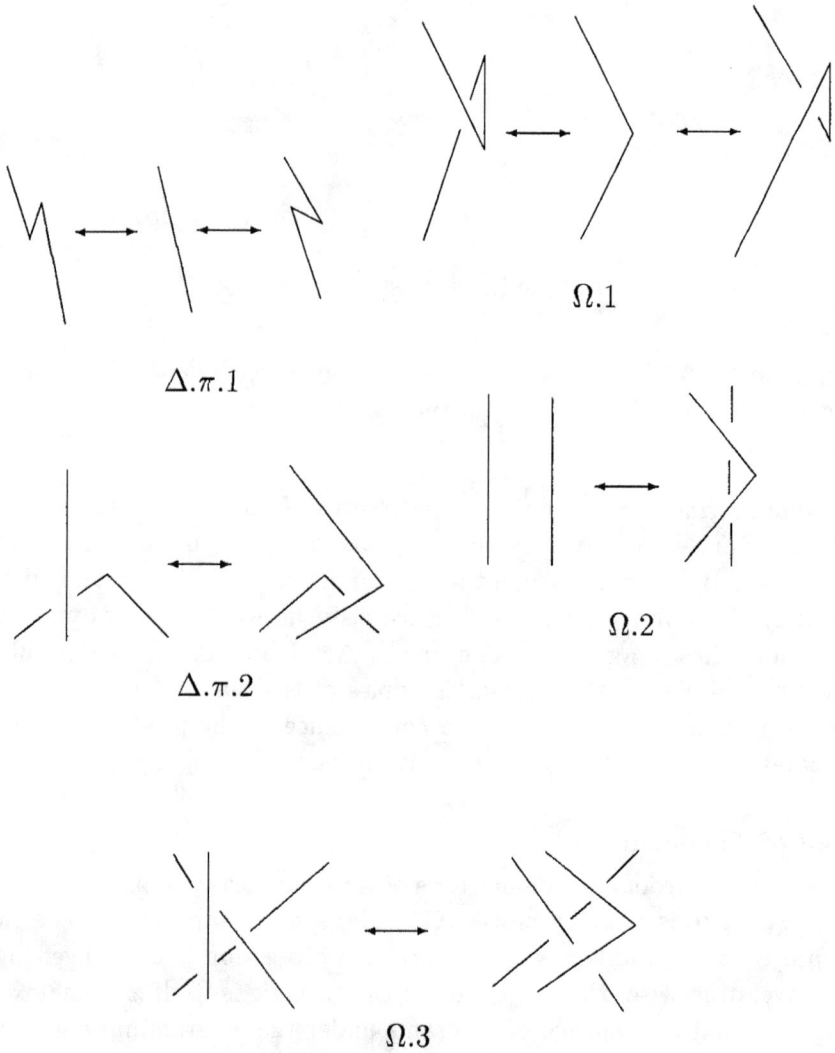

Figure 2.3: Reidemeister's Moves

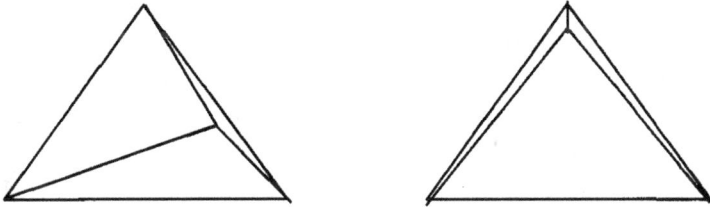

Figure 2.4: Subdivisions Useful to Avoid Non-regular Projections

move across a triangle parallel to the plane of projection.

For the converse, we would like to proceed by simply considering the effect of isotopy by a single move across a triangle on the projection. However, before doing so, we must show that we may assume, without loss of generality, that each move not only begins, but ends, with a regular projection.

Now, if we subdivide any triangle into smaller triangles, the move across the triangle can be realized instead as a sequence of moves across the smaller triangles. This observation is the key both to the remainder of the proof, and to solving the difficulty just mentioned.

If a move results in a non-regular projection, we can replace it with three moves across smaller triangles as in Figure 2.4. The subdivision point must be chosen so that the move across the large triangle(s) results in regular projections, and near enough to the new arc. Near enough, here, means

1. within a neighborhood bounding the new arc away from the triangles of later moves, if the non-regularity is removed by moves not involving the new arc, or

2. so that the convex hull of the triangle of the move removing the non-regularity and the image of the nearest-neighbor projection of its starting arc across the thin triangle(s) does not intersect the remainder of the link, if the non-regularity is removed by a move involving the new arc.

In either case, we replace the sequence of moves with a sequence in which the move introducing the non-regularity is replaced by the move(s) across the large triangle(s). In the first case, the move(s) across the small triangle(s) is (are) made just after the move which removed the non-regularity in the original sequence. In the second case, the move which removed the non-regularity in the original sequence is replace by moves across the other faces of the convex hull of item 2, and subdivisions of the thin triangle(s).

Now, we may assume that all of our moves begin and end with links whose projection onto a given plane are regular. Let the complexity of a move to be given by the number of edges, vertices and crossings of the link whose projection intersect the projection of the interior triangle of the move. If the move has a complexity greater than three, or if there are no vertices or crossings whose projection lies in the interior of the triangle and the move has a complexity greater than one, we can replace the move with a sequence of less complex moves across a subdivision of the triangle.

It therefore suffices to show that the result holds for moves of minimal complexity: those of complexity 0 and 1 with no vertices or crossings in the projection of the interior of the triangle, and those of complexity 3 involving a vertex or crossing.

Now, a move of complexity 0 is immediately seen to be one of type $\Delta.\pi.1$. A move of complexity 1 is of type $\Omega.1$ in the case where the edge whose projection is interior to the projected triangle is incident with the arc being moved, of type $\Delta.\pi.2$ in the case where it crosses the arc being moved on the boundary, and of type $\Omega.2$ otherwise.

A move of complexity 3 involving a vertex is of type $\Omega.2$ if the arc including the vertex does not cross the edge being moved, and of type $\Delta.\pi.2$ if it does.

Finally, a move of complexity 3 involving a crossing is plainly of type $\Omega.3$. $\square$

It is easy to incorporate orientation data into a knot diagram: one need only equip the projection of each component of the link with an arrow on one of its arcs

Using crossing signs, itt is now possible to give a combinatorial definition of linking number:

Definition 2.14 *Given two components* K_1, K_2 *of a link L, the* linking number $lk(K_1, K_2)$ *is* $\frac{1}{2}(c_+ - c_-)$, *where* c_+ *(resp.* c_-*) is the number of positive (resp. negative) crossings involving one arc of* K_1 *and one arc of* K_2 *in some diagram of the link.*

It can be easily verified that this number is invariant under the Reidemeister moves, and is thus independent of the choice of diagram.

What is slightly less clear is that one can incorporate the framing information for an oriented framed link in the knot diagram as well: perform an ambient isotopy which is trivial outside of a tubular neighborhood of the link to make the ribbon parallel to the plane of projection, and pointing right with respect to the orientation vectors. In doing this, one may have to introduce kinks into the diagram (by moves of the form $\Omega.1$).

The ambient isotopy class of the oriented framed link can then be recovered from the resulting knot diagram by mapping the ribbon in such a way that it lies to the right of the curve when traversing it in the direction determined by the orientation. The framing determined in this way from a diagram is called the *blackboard framing* (cf. [36]). This process of introducing kinks to "flatten" the ribbon makes clear that the move $\Omega.1$ does not preserve the ambient isotopy type of the framed link which is recovered from the diagram.

All of the other Reidemeister moves may readily be seen to preserve the equivalence class of oriented links with the blackboard framing. Omitting $\Omega.1$ from the Reidemeister moves give a combinatorial notion of equivalence called *"regular isotopy"* which was used by Kauffman [32] in his formulation of the Jones polynomial, the so-called "Kauffman bracket" (cf. also [29]).

For our purposes, this combinatorial notion is less useful than a reduction to diagrams of ambient isotopy of framed oriented links. For this, we need to replace $\Omega.1$ with a substitute move which does respect the framing. To do this, we need to examine how the various cases of $\Omega.1$ change the blackboard framing. Observe that those which introduce

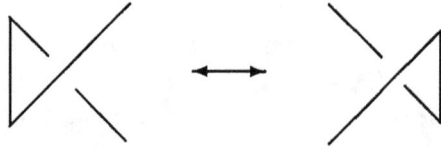

Figure 2.5: The Framed First Reidemeister Move

positive crossings change the framing (thought of as an integer) by $+1$, while those with negative crossings change it by -1. It therefore follows that any combinations of moves of type $\Omega.1$ which change the framing by 0 must be admitted as moves.

Now any such sequence of Reidemeister moves which preserves the framing can be modified by moves of the types other than $\Omega.1$ (by sliding curls along the component of the link) in such a way that moves of type $\Omega.1$ which increase the framing are paired with moves of type $\Omega.1$ which decrease the framing in small balls (or disks in the projection). By use of the simplest "Whitney trick"—the fact that moves of types $\Omega.2$ and $\Omega.3$ suffice to remove a pair of loops, provided they have opposite crossings, and lie on opposite sides of the arc in the projection, all of the various cases can be reduced to the single move in Figure 2.5.

2.2   Categories, Functors, Natural Transformations

We now turn to the basic notions from category theory needed for this study. The reader interested in a more thorough treatment is referred to Mac Lane [40], which contains most of the standard elementary definitions and theorems. We repeat those of particular importance for this study in this section and the next chapter.

Definition 2.15 (objects-and-arrows) *A category C consists of two collections $Ob(C)$ and $Arr(C)$, whose elements are called, respectively, the* objects *and* arrows *of C together with assignments of objects* target(f) *and* source(f) *to each arrow f; of an arrow Id_X to each object X; and* *of an arrow denoted fg or $g(f)$, called the* composition *of f and g, to each pair of arrows f, g for which* target$(f) =$ source(g), *and satisfying*

$$
\begin{aligned}
\text{source}(Id_X) &= X \\
\text{target}(Id_X) &= X \\
Id_{\text{source}(f)}f &= f \\
fId_{\text{target}(f)} &= f \\
h(g(f)) &= h(g)(f).
\end{aligned}
$$

The arrows of a category are also, particularly in concrete settings, referred to at *morphisms* or *maps*. At first, we will adhere to calling them "arrows", but as we move to setting where the other names are common, we will begin to use them interchangeably.

The coyness of not describing source$(-)$, target$(-)$, Id_- and composition as functions is traditional (and to some minds necessary) because the collections involved are often proper classes.[2] The reader who dislikes bothering about the niceties of set theory may proceed safely: all of the categories which will occur in this book, outside of some illustrative examples in this section, are either *small* (that is, both $Ob(\mathcal{C})$ and $Arr(\mathcal{C})$ are sets) or *essentially small* (that is *equivalent*–as defined below–to a small category). One other notion connected with size in the set-theoretic sense should be mentioned: if for every pair of objects X, Y, the collection of arrows with source$(f) = X$ and target$(f) = Y$ is a set, we say the category is *locally small*. All categories considered herein are locally small.

<hr/>

[2]Many categorists object to the habit of mind which tries to place all of mathematics on a set-theoretic foundation. After all, when is the last time anyone ever actually cared about the ϵ-tree defining an element of a smooth manifold? For the insistently set-theory minded, we will dispense with the problems usually raised concerning sizes of categories by using a sufficiently strong large-cardinal axiom—Grothenieck's Axiom of Universe. Those size problems which do not collapse in the face of this axiom, and there are some, do not arise in this study.

Definition 2.16 *For a locally small category* C, *the set of all arrows with* source$(f) = X$ *and* target$(f) = Y$ *is denoted* $Hom_C(X, Y)$ *or simply* $C(X, Y)$, *and called the* hom-set *from* X *to* Y.

If we do write the structure given in Definition 2.15 in terms of sets and functions we have

$$\text{source} : Arr(C) \longrightarrow Ob(C)$$

$$\text{target} : Arr(C) \longrightarrow Ob(C)$$

$$Id : Ob(C) \longrightarrow Arr(C)$$

$$-(-) : Arr(C) \times_{Ob(C)} Arr(C) \longrightarrow Ob(C)$$

satisfying the functional equations given element-wise in the definition.

It will be observed that both source and target split Id, and thus Id is a bijection between $Ob(C)$ and its image. As we are concerned only with the structure of the category, not with the identity of its objects or arrow in some external ideal universe, this bijection allows us to forget the objects entirely: we can consider the identity maps themselves as the objects. Doing so gives an alternative definition of category which is sometimes more convenient:

Definition 2.17 (arrows-only) *A category* C *is a collection* C *whose elements are called "arrows", equipped with two unary operations* **source** *and* **target** *and a partially defined operation denoted by the null infix, with the property that* fg *is defined if and only if* **target**$(f) = $ **source**(g), *and satisfying*

$$\begin{aligned}
\textbf{source}(\textbf{source}(f)) &= \textbf{source}(f) \\
\textbf{target}(\textbf{source}(f)) &= \textbf{source}(f) \\
\textbf{source}(\textbf{target}(f)) &= \textbf{target}(f) \\
\textbf{target}(\textbf{target}(f)) &= \textbf{target}(f) \\
\textbf{source}(f)f &= f
\end{aligned}$$

$$f\,\mathbf{target}(f) = f$$
$$[fg]h = f[gh]$$

Example 2.18 Sets: *Objects are all sets in your favorite model of your favorite set-theory; arrows are all set-functions; source is domain; target is codomain; Id_X for any set X is the identity function on X; and composition is composition of set-functions.*

Example 2.19 Esp: *Objects are all topological spaces; arrows are all continuous maps; source is domain; target is codomain; Id_X for any space X is the identity function on X; and composition is composition of continuous maps.*

Example 2.20 K − mod: *Fix a ring K. Objects are all K modules; arrows are all K-linear maps; source is domain; target is codomain; Id_X is the identity map on X; and composition is composition of K-linear maps.*

Examples of this sort can be multiplied *ad infinitum*: take as objects all examples of some mathematical structure, and as arrows all maps preserving (some part of) the structure, ... In these cases it is most convenient to use the objects-and-arrows definition. This is not always the case. Consider

Example 2.21 G: *Fix a group (or monoid) G. Consider its elements as arrows with composition defined by the group law, and* **source** *and* **target** *given by the the constant map to e, the identity element.*

More important for this study are:

Example 2.22 Tang *(resp.* **Otang, Frtang***): Consider as arrows all equivalence classes of tangles (resp. oriented tangles, framed oriented*

tangles). **source**(T) *(resp.* **target**$(T))$ *is the linear embedding of a disjoint union of copies of* $\mathbb{I}$ *which is constant in the first two coördinates and intersects* T *at each point of* $\mathbb{I}^2 \times \{0\}$ *(resp.* $\mathbb{I}^2 \times \{1\})$ *in the same set of points as* T *with (resp. the same set with the same orientation, the same set with the same orientation and framing). The composition of two tangles* T_1, T_2 *has as underlying 1-manifold the union of the underlying 1-manifolds of* T_1 *and* T_2 *with the points of the boundary lying in the face containing the common source/target identified. The composition* $T_1 T_2$ *is then defined by the map on this underlying 1-manifold given as a composition of* $T_1 \coprod T_2$, *with the map* $\gamma_3 : \mathbb{I}^3 \coprod \mathbb{I}^3 \to \mathbb{I}^3$ *given by*

$$(x, y, z) \mapsto (x, y, \frac{z}{2}) \text{ for elements of the first summand}$$

$$(x, y, z) \mapsto (x, y, \frac{z+1}{2}) \text{ for elements of the second summand,}$$

with the connected components PL homeomorphic to $\mathbb{I}$ *reparameterized to preserve the condition at the boundary.*

It requires a little work to verify that this actually gives rise to a category. The conditions involving only **source** and **target**, but not composition, are immediate. To verify the other conditions, observe first that the two sides of the equations are certainly not equal by construction until we pass to equivalence classes. It is necessary to construct a PL (smooth) ambient isotopy rel boundary to verify the equations.

The required isotopies are constant in the first two coördinates of $\mathbb{I}^3$ and in all coördinates in a neighborhood of $\partial \mathbb{I}^3$. In the third coördinate they are given in a set F of the form $[\epsilon, 1 - \epsilon]^3$ by (smoothings of) the PL maps shown schematically in Figure 2.6. The extension of this isotopy given by Lemma 2.5 then gives an isotopy which preserves the condition on the boundary.

Example 2.23 n-**Cobord**: *As objects, take oriented smooth* $(n-1)$-*manifolds. As arrows, let* $Hom_{n-\textbf{Cobord}}(M, N)$ *be the set of all equivalence classes of oriented* n-*manifolds with boundary* X *equipped with*

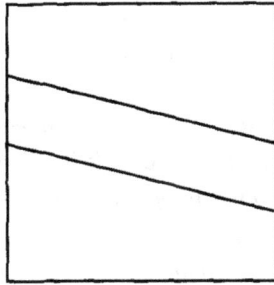

associativity

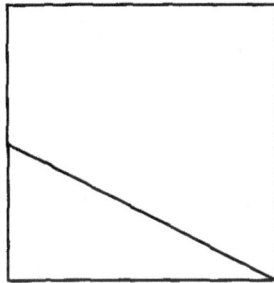

right identity

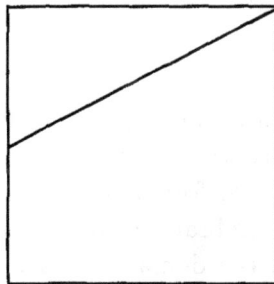

left identity

Figure 2.6: Isotopies Giving Identity and Associativity Conditions in Categories of Tangles

a diffeomorphism $\phi : -M \coprod N \to \partial X$, where $-M$ denotes M with its orientation reversed, and where (X, ϕ) is equivalent to (Y, ψ) when there exists a diffeomorphism $\Theta : X \to Y$ such that $\Theta(\phi) = \psi$.

Composition is given by "gluing" the source of one cobordism to the target of another and giving the resulting manifold the unique smooth structure for which charts in the interior of each manifold are charts and the bicollar neighborhood of the "gluing locus" obtained by gluing collar neighborhoods of the boundary components has the product smooth structure. It is clear that $Id_N = N \times \mathbb{I}$ with the obvious diffeomorphism of $-N \coprod N$ with $\partial(N \times \mathbb{I})$ as structure map.

A number of auxiliary notions arise almost immediately from the definition of categories. Those needed in this study include

Definition 2.24 *An arrow f in a category is an* isomorphism *if it is invertible in the sense that there exists an arrow g such that*

$$\textbf{source}(f) = \textbf{target}(g) \text{ and } \textbf{target}(f) = \textbf{source}(g),$$

and satisfying $fg = \textbf{source}(f)$ and $gf = \textbf{target}(f)$.

The reader will recall that in objects-and-arrows terminology

$$\textbf{source}(f)$$

is the identity arrow on source(f), and similarly for targets.

For the purposes of this study, it is important to observe that the axioms of categories (in either formulation) are axioms of what Freyd has called an "essentially algebraic theory": the operations can be ordered in such a way that the domain of each operation is described by equations in earlier operations. (For objects-and-arrows, one has a two-sorted theory, while for arrow-only one has a one-sorted theory.)

Thought of in this way, there are a number of constructions of categories which immediately present themselves. One can present categories by generators and relations (for example, the category generated by a single arrow f subject to the relation $\textbf{source}(f) = \textbf{target}(f)$ is the additive monoid of ℕ regarded as a category). It is also clear that the disjoint union of two (or an indexed family of) categories is again a category in an obvious way, and that the cartesian product of two (or an indexed family of) categories is a category with component-wise operations (either use the arrows-only formalism, or take disjoint unions or products of the sets of objects and of arrow separately.[3]). Likewise, given a category one can define subcategories as subsets of the arrows closed under the operations.

One particular type of subcategory bears mention:

Definition 2.25 *Given a set of objects (or equivalently* **source** *arrows)* S *in a category* C, *the* full subcategory on S, $full(S)$, *is the subcategory of* C *consisting of all arrows whose source and target both are elements of* S.

Another construction of one category from a given category is presented by the fact that the axioms admit a symmetry: reverse source and target, and reverse the order of composition. Given a category C, the category with the same arrows, but with the source and target operations switched and the order of composition reversed is called the *opposite category,* and is denoted C^{op}.

There is then an obvious notion of homomorphism of categories:

Definition 2.26 (arrows-only) *A functor* F *from a category* C *to a category* D *(denoted* $F : C \to D$*) is an assignment of an arrow* $F(f)$ *of* D *to each arrow* f *of* C, *which preserves* **source**, **target**, *and composition in the sense that*

$$\mathbf{source}(F(f)) = F(\mathbf{source}(f))$$
$$\mathbf{target}(F(f)) = F(\mathbf{target}(f))$$
$$F(fg) = F(f)F(g)$$

---

[3] Again, for folks who worry about set-theoretic size issues, we will use the Axiom of Universe to side-step the issue, and use *sets* of objects and arrows, albeit in a larger model of set-theory.

Of course, this provides another example of categories: **Cat**, which has as objects all (small) categories, and as arrows all functors, with the obvious source, target, identity, and composition operations.

Example 2.27 Underlying or "Forgetful" functors *Consider our abundant supply of categories: sets equipped with some structure as objects, and maps which preserve the structure as arrows (e.g., rings and ring-homomorphisms). Now, consider only part of the structure (e.g. additive abelian group and abelian group homomorphisms). There is then a functor which takes each object to itself (with some of its structure forgotten) and each structure-preserving map to the same map (which necessarily preserves the part of the structure not forgotten). Thus, "additive group of" may be thought of as a functor from the category of rings to the category of abelian groups.*

If one forgets all of the structure, one obtains the "underlying set" functor.

Example 2.28 Group homomorphisms *Considering groups G and H as categories, functors from G to H are precisely group homomorphisms from G to H.*

Example 2.29 Structure functors for disjoint unions *Given an indexed family of categories $\{C_j\}_{j \in J}$, the disjoint union $\coprod_{j \in J} C_j$ is a category with the obvious source, target, identities and compositions. The inclusions $\iota_j : C_j \to \coprod_{j \in J} C_j$ are functors. Likewise, given functors $F_j : C_j \to \mathcal{D}$, there is a (unique) functor $F : \coprod_{j \in J} C_j \to \mathcal{D}$ such that $F(\iota_j) = F_j$ for all j.*

Example 2.30 Structure functors for products *Given an indexed family of categories $\{C_j\}_{j \in J}$, the product $\prod_{j \in J} C_j$ is a category with componentwise source, target, identities and compositions. The projections $\pi_j : \prod_{j \in J} C_j \to C_j$ are functors. Likewise, given functors $F_j : \mathcal{D} \to C_j$, there is a (unique) functor $F : \mathcal{D} \to \prod_{j \in J} C_j$ such that $\pi_j(F) = F_j$ for all j.*

Example 2.31 (Co)homology groups *Consider the category* **Esp**, *with topological spaces as objects, and continuous functions as arrows. Then the assignments* $U \mapsto H_i(U)$ $(i^{th}$ *homology group), (resp.* $U \mapsto H^i(U)$ $(i^{th}$ *cohomology group)) have corresponding assignments on arrows which give functors from* **Esp** *to* **Ab**, *the category of abelian groups and group homomorphisms (resp.* **Ab**^{op}, *its opposite category).*

Example 2.32 (Co)homology groups *Consider the category* **Esp**$_2$, *with pairs* $U \supset V$ *of topological spaces as objects, and continuous functions on the larger space which preserve the smaller space as arrows. Then the assignments* $(U \supset V) \mapsto H_i(U, V)$ *(resp.* $(U \supset V) \mapsto H^i(U, V)$*) have corresponding assignments on arrows which give functors from* **Esp**$_2$ *to* **Ab**, *the category of abelian groups and group homomorphisms (resp.* **Ab**^{op}, *its opposite category).*

The cohomological cases of the last two examples are often phrased in terms of a "contravariant functor". As we will have cause to use this notion, we make

Definition 2.33 *A* contravariant functor from $\mathcal{C}$ to $\mathcal{D}$ *is a functor from* $\mathcal{C}$ *to* $\mathcal{D}^{op}$.

Once it is observed that $[\mathcal{X}^{op}]^{op} = \mathcal{X}$, and that the same data which describes a functor from $\mathcal{C}$ to $\mathcal{D}$ describes a functor from $\mathcal{C}^{op}$ to $\mathcal{D}^{op}$, we see that a contravariant functor from $\mathcal{C}$ to $\mathcal{D}$ can equally well be regarded as a functor from $\mathcal{C}^{op}$ to $\mathcal{D}$.

One of the remarkable features of category theory is that, as in homotopy theory, there is a good notion of maps between maps:

Definition 2.34 *Let* F *and* G *be functors from* $\mathcal{C}$ *to* $\mathcal{D}$. *Then a* natural transformation ϕ *from* F *to* G *(denoted* $\phi : F \Rightarrow G$*) is an assignment to each object* X *of* $\mathcal{C}$ *of an arrow* $\phi_X : F(X) \to G(X)$ *of* $\mathcal{D}$, *satisfying*

$$\phi_X G(f) = F(f)\phi_Y$$

for every arrow $f : X \to Y$ *of* $\mathcal{C}$.

It is easy to see that if F, G and H are functors from $\mathcal{C}$ to $\mathcal{D}$, and $\phi : F \Rightarrow G$ and $\psi : G \Rightarrow H$ are natural transformations, then $\phi\psi$, given by $[\phi\psi]_X = \phi_X\psi_X$, is a natural transformation from F to H. Thus, we have not merely a hom-set of functors from $\mathcal{C}$ to $\mathcal{D}$, but a "hom-category" of functors from $\mathcal{C}$ to $\mathcal{D}$ and natural transformations between them.

This is the first example of something quite important to this study: categories in which the hom-sets have an additional structure. We will take up this notion in Chapter 10.

Example 2.35 *If two group homomorphisms $f, g : G \to H$ are conjugate, the assignment of the conjugating element (arrow) in H to the unique object of G is a natural transformation from f to g (regarded as functors).*

Example 2.36 *Consider the category $\mathbf{Esp}_2$ of Example 2.32. There are two obvious functors $P_1, P_2 : \mathbf{Esp}_2 \to \mathbf{Esp}$ which assign to the pair the ambient space and the subspace, respectively.*

The assignment to each pair of the map induced on homology by the inclusion is a natural transformation from $H_i(P_2)$ to $H_i(P_1)$.

Similarly, assignment to each pair (U, V) of the canonical map from $H_i(U)$ to $H_i(U, V)$ is a natural transformation from $H_i(P_1)$ to H_i (the latter referring to relative homology).

Finally, the assignment to pairs (U, V) of the connecting homomorphisms $\phi : H_i(U, V) \to H_{i-1}(V)$ is a natural transformation from H_i to $H_{i-1}(P_2)$.

When every component arrow of a natural transformation is of a particular type (e.g., an isomorphism) we refer to the natural transformation as a *"natural <name of type of arrow>"* (e.g., a natural isomorphism).

The analogy with homotopy theory suggests the following analogue of homotopy equivalence (though it can be motivated by other purely category-theoretic considerations):

Definition 2.37 *An* equivalence of categories *between C and D is a pair of functors $F : C \to D$, $G : D \to C$ and a pair of natural isomorphisms $\phi : FG \Rightarrow Id_C$ and $\psi : GF \Rightarrow Id_D$.*

By abuse of language we say a functor $F : C \to D$ is an equivalence of categories if there exist G, ϕ and ψ as above. We say two categories C and D are equivalent *if there exists an equivalence of categories between them.*

It is an easy exercise (which involves the behavior of natural transformations under composition of functors) to see that equivalence of categories is indeed an equivalence relation.

Finally, we mention one of the more pleasant features of category theory: we can summarize equations between various compositions of arrows by using diagrams of nodes labeled by objects of the category joined by oriented edges (arrows!) labeled by arrows of the category. Such a diagram is said to *commute* if for every pair of paths along the arrows of the diagram with the same starting and ending points, the (iterated) compositions of the arrows along the two paths are equal.

We will follow the convention that the presentation of a diagram as if it were a statement asserts that the diagram commutes, and the presentation of a diagram with variables representing objects (or maps) as if it were a statement asserts that all instantiations of the diagram commute. For example, the definition of a natural transformation may be rephrased as follows:

A natural transformation *between the functors $F, G : C \to D$ is an assignment to each object X of C of an arrow $\phi_X : F(X) \to G(X)$ such that the condition of Figure 2.7 holds.*

$$F(X) \xrightarrow{\;F(f)\;} F(Y)$$

$$\phi_X \downarrow \qquad\qquad\qquad \downarrow \phi_Y$$

$$G(X) \xrightarrow[\;G(f)\;]{} G(Y)$$

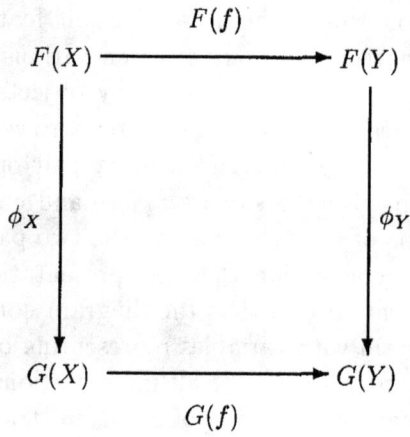

Figure 2.7: A Naturality Square

Chapter 3

Monoidal Categories, Functors and Natural Transformations

We now turn to the consideration of the most elementary natural examples of "categories with structure." That is, we will now study categories equipped with additional operations given by functors or natural transformations, and satisfying additional axioms in terms of these; and those functors and natural transformations which respect, or are adapted to the structures in some reasonable way.

We refer to the components of any natural transformations in the specification of a type of "category with structure" (or "functor with structure") as *structure maps*. Thus far we have scrupulously observed the use of "arrow" as the sole name for the elements of a category (in the arrows-only formalism). As we pass to more and more engagement with concrete examples, we will more and more use the suggestive names of "morphism" or "map" to refer to arrows in a category.

Definition 3.1 *A* monoidal category $\mathcal{C}$ *is a category* $\mathcal{C}$ *equipped with a functor* $\otimes : \mathcal{C} \times \mathcal{C} \to \mathcal{C}$ *and an object* I, *together with natural isomorphisms* $\alpha : \otimes(\otimes \times 1_{\mathcal{C}}) \Rightarrow \otimes(1_{\mathcal{C}} \times \otimes)$, $\rho : \otimes I \Rightarrow 1_{\mathcal{C}}$ *and* $\lambda : I\otimes \Rightarrow 1_{\mathcal{C}}$,

*satisfying the pentagon and triangle coherence conditions of Figure 3.1
and the bigon ($\rho_I = \lambda_I$) coherence condition   (cf. [40]).[1] Similarly, a
semigroupal category is a category equipped with only $\otimes$ and α, satisfying the pentagon of Figure 3.1. A monoidal (semigroupal) category is
strict   if all of its structure maps are identity maps.*

Monoidal categories are quite common "in nature" as the next few
examples will show, and, of course, any monoidal category is *a fortiori*
a semigroupal category. The weaker notion is defined principally as a
convenience for our later examinations of deformation theory.

Example 3.2 (Sets, $\times$, $\{*\}$, α, ρ, λ) *is a monoidal category, where $\times$ is
cartesian product, $\alpha_{A,B,C}$ is given by $((a,b),c) \mapsto (a,(b,c))$, ρ_A is given
by $(a,*) \mapsto a$ and λ_A is given by $(*,a) \mapsto a$.*

The coherence conditions (pentagon and triangles) can readily be
verified for Example 3.2. For example, the pentagon is given by

$$(((a,b),c),d) \mapsto ((a,b),(c,d)) \mapsto (a,(b,(c,d))) =$$
$$(((a,b),c),d) \mapsto ((a,(b,d)),d) \mapsto (a,((b,c),d)) \mapsto (a,(b,(c,d))).$$

Similarly, the naturality of the structure maps is easy to verify. For
example, in the case of ρ, if $f : A \to B$ is any set function, then the
two legs of the naturality square are given on elements by $(a,*) \mapsto
(f(a),*) \mapsto f(a)$ and $(a,*) \mapsto a \mapsto f(a)$.

This example may be extended to similar examples for any mathematical structure which can be defined component-wise (e.g. topological
spaces, groups, smooth manifolds, rings) and the appropriate maps all
admit "cartesian" monoidal structures.

Example 3.3 (Sets, $\coprod$, $\emptyset$, α, ρ, λ) *is a monoidal category, where $\coprod$ is
disjoint union, and α, ρ and λ all map elements to themselves, regarded
as elements in the "other" set.*

---

[1]It can be shown that the bigon condition is redundant, but we leave it as an
exercise to the reader to show that we can "Let bigons be bygones."

$$(A \otimes B) \otimes (C \otimes D)$$

α α

$$((A \otimes B) \otimes C) \otimes D \qquad\qquad\qquad A \otimes (B \otimes C) \otimes D))$$

$\alpha \otimes D$ $A \otimes \alpha$

$$(A \otimes (B \otimes C)) \otimes D \xrightarrow{\quad\alpha\quad} A \otimes ((B \otimes C) \otimes D)$$

$$(A \otimes I) \otimes B \xrightarrow{\quad\alpha\quad} A \otimes (I \otimes B)$$

$\rho \otimes B$ $A \otimes \lambda$

$$A \otimes B$$

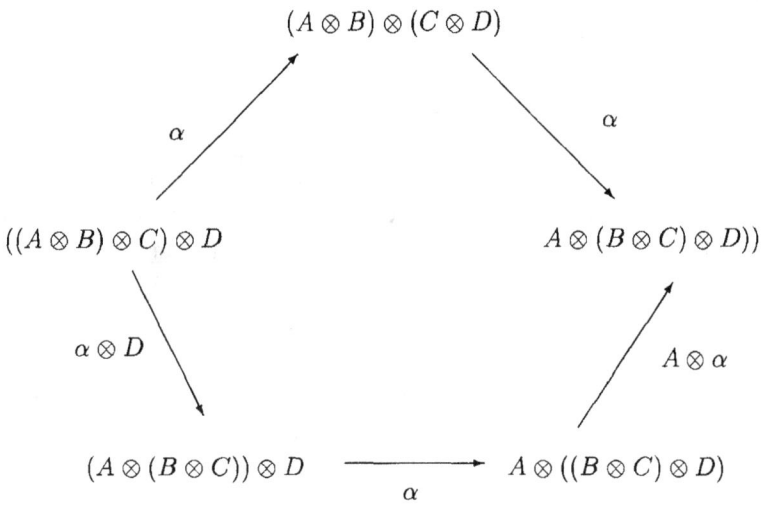

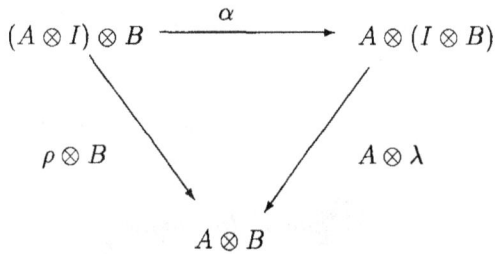

Figure 3.1: Coherence Conditions for Monoidal Categories

In this case, there is nothing to check to see that the coherence conditions hold, while naturality is almost as immediate.

Again, this example will extend to give similar examples for any mathematical structure which is given trivially on a disjoint union in terms of the structures on the summand (e.g., topological spaces, smooth manifolds). Similar examples exist for any type of algebraic structure which can be given by generators and relations: the monoidal product is given by using the disjoint union of the underlying sets of two objects as a set of generators and imposing the relations which exist in the summands. As a particular case, we have:

Example 3.4 (**Grps**, $*, 1, \alpha, \rho, \lambda$) *is a monoidal category, where $*$ is the free product, 1 is the trivial group, and α, ρ and λ are given on generators (elements of the groups whose free product is taken) by the map of the same name in the previous example targeted at the set of generators.*

The next two examples are of particular importance to us:

Example 3.5 *Let K be a field. Then*

$$(K-\mathbf{v.s.}, \otimes, \alpha, \rho, \lambda)$$

and

$$(K-\mathbf{v.s.}_{f.d.}, \otimes_K, K, \alpha, \rho, \lambda),$$

are monoidal categories, where $K-\mathbf{v.s.}$ and $K-\mathbf{v.s.}_{f.d}$ are the categories of K-vectorspaces and K-linear maps, and of finite dimensional K-vectorspaces and K-linear maps, respectively, and where $\otimes_K$ is the tensor product over K (universal target for bilinear maps), $\alpha_{U,V,W}$ is given on the usual spanning set by $(u \otimes v) \otimes w \mapsto u \otimes (v \otimes w)$ and ρ and λ are right and left scalar multiplication, respectively.

Since to check the equality of linear maps, it suffices to check them on a spanning set, the verification that the associativity is coherent in Example 3.5 is given by a calculation formally identical to that for Example 3.2, but applied to parenthesized iterated tensor products of elements rather than parenthesized n-tuples of elements.

Definition 3.6 *A bialgebra over a field K is an algebra A, equipped with algebra homomorphisms $\Delta : A \otimes_K A \to A$ and $\epsilon : A \to K$, which satisfy the conditions*

$$\alpha_{A,A,A}([\Delta \otimes A](\Delta)) = [A \otimes \Delta](\Delta) \quad (coassociativity),$$

$$\lambda([\epsilon \otimes A](\Delta)) = Id_A \quad (left\ counitalness),$$

and

$$\rho([A \times \epsilon](\Delta)) = Id_A \quad (right\ counitalness).$$

(Observe that $A \otimes_K A$ is a K-algebra with multiplication given on a basis by $[x \otimes y] \cdot [w \otimes z] = x \cdot w \otimes y \cdot z$.)

Aside: we include α, λ and ρ, which are usually suppressed, because we have not yet introduced the theorem which justifies their exclusion.

Example 3.7 *Let A be a bialgebra over a field K with*

$$\Delta(a) = \sum_{i=1}^{n(a)} a_i' \otimes a_i'' .$$

Then $(A - \mathbf{mod}, \otimes_K, K, \alpha, \rho, \lambda)$, is a monoidal category, where $A - \mathbf{mod}$ is the category of K-finite dimensional A-modules, and the structure maps are as in the previous example, with $X \otimes_K Y$ given an A-module structure by

$$a \cdot (x \otimes y) = \sum_{i=1}^{n(a)} a_i' \cdot s \otimes a_i'' \cdot y$$

and K given an A-module structue by $a \cdot k = \epsilon(a)k$, where the null infix is multiplication in K.

The important thing to observe is that the structure maps of Example 3.5 are necessarily module homomorphisms when tensor products and K are given A-module structures as specified. Coherence and naturality are then simply restrictions of the corresponding properties in K − **v.s.**. In Chapter 4 we provide numerous examples of bialgebras.

Example 3.8 Tang *(resp.* **Otang, Frtang***) is a monoidal category when equipped with the empty tangle as I and* $\otimes$ *given as follows: for any two tangles* T_1, T_2, $T_1 \otimes T_2$ *has as underlying 1-manifold the disjoint union of the underlying 1-manifolds of* T_1 *and* T_2. $T_1 \otimes T_2$ *is then the mapping of this 1-manifold the composition of* $T_1 \coprod T_2$ *with the map* $\gamma_2 : \mathbb{I}^3 \coprod \mathbb{I}^3 \to \mathbb{I}^3$ *given by*

$$(x, y, z) \mapsto (\frac{x}{2}, y, z) \text{ for elements of the first summand, and}$$

$$(x, y, z) \mapsto (\frac{x+1}{2}, y, z) \text{ for elements of the second summand.}$$

The structure of this example will considered in detail in Chapter 7.

Example 3.9 *n*-cobord *is a monoidal category with* $\otimes$ *given on both objects and maps by disjoint union and with the empty* $(n-1)$-manifold *as I. The structure maps are given by trivial cobordisms (products with* $\mathbb{I}$*) with attaching maps induced by the identity at the source, and the coherence map from the monoidal structure of Example 3.3 at the target.*

One might wonder why two other obvious triangles relating the associator α and the unit transformations λ and ρ — namely,

$$\alpha_{A,B,I} Id_A \otimes \rho_B = \rho_{A \otimes B}$$

and

$$\alpha_{I,B,C} \lambda_{B \otimes C} = \lambda_B \otimes C$$

— are omitted from the definition. It turns out they are a consequence of the given coherence conditions. We will later need:

Lemma 3.10 *In any monoidal category, and for any objects A, B, and C, the equations $\alpha_{A,B,I} Id_A \otimes \rho_B = \rho_{A \otimes B}$ and $\alpha_{I,B,C} \lambda_{B \otimes C} = \lambda_B \otimes C$ hold.*

proof: We give the proof of the first only. The other follows from the same proof applied to the monoidal category obtained by reversing the order of $\otimes$, inverting α to obtain an new associator, and reversing the roles of λ and ρ.

Consider the case of the pentagon with initial vertex

$$[[[A \otimes B] \otimes I] \otimes I].$$

This pentagon can be filled with

- two naturality squares for α, one including the middle arrow of the three-arrow side of the pentagon, the other including the second arrow of the two-arrow side, and both having $\alpha_{A,B,I}$ as the other instance of α parallel to the given arrow in the pentagon,

- an instance of the triangle coherence condition for the objects $A \otimes B$ and I containing the first edge of the two-arrow side of the pentagon, and one edge of the second naturality square above,

- a prolongation of the triangle coherence condition whose edges are formed by the third-arrow of the three arrow side of the pentagon and one arrow from each of the naturality squares, and

- a prolongation of the desired triangle by right monoidal product with Id_I.

It then follows (once suitable maps have been inverted) that the pentagon, the two naturality squares and the other two triangles just described give a commutative filling of the prolongation of the desired triangle.

The desired triangle can then be shown to commute by filling it with the prolongation just shown to commute and three naturality squares for ρ^{-1}.

The reader is encouraged to write out all of the diagram fillings just described. □

Because categorical structures can be preserved "on the nose", preserved up to (natural) isomorphism, or even up to a natural transformation in one direction or the other, there are a variety of different types of functors "preserving" a monoidal structure:

Definition 3.11 *A* lax monoidal functor $F : C \to D$ *between two monoidal categories* C *and* D *is a functor* F *between the underlying categories, equipped with a natural transformation*

$$\tilde{F} : F(-) \otimes F(-) \to F(- \otimes -)$$

and a map $F_I : I \to F(I)$, *satisfying the hexagon and two squares of Figure 3.2.*

An oplax monoidal functor $F : C \to D$ *between two monoidal categories* C *and* D *is a functor* F *between the underlying categories, equipped with a natural transformation*

$$\tilde{F} : F(- \otimes -) \to F(-) \otimes F(-)$$

and a map $F_0 : F(I) \to I$, *satisfying the hexagon and two squares of Figure 3.3.*

A strong monoidal functor $F : C \to D$ *between two monoidal categories* C *and* D *is a functor* F *between the underlying categories, equipped with a natural isomorphism*

$$\tilde{F} : F(- \otimes -) \to F(-) \otimes F(-)$$

and an isomorphism $F_0 : F(I) \to I$, *satisfying the hexagon and two squares of Figure 3.3.*

A strict *monoidal functor is a strong monoidal functor for which all components of* $\tilde{F}$ *and* F_0 *are identity maps .*

Lax, oplax, strong and strict semigroupal functors are defined similarly.

We refer to the components of the natural transformations and maps specified in these definitions, and to their inverses (if any), as *structure maps*. Likewise, a map which is obtained from some other map f by forming an iterated monoidal product of f with identity maps for various objects is called a *prolongation* of f. Sometimes by abuse of terminology prolongations of structure maps are themselves refered to as structure maps.

We will also refer to a diagram obtained by applying the same iterated monoidal product with identity maps to every map of a given diagram as a prolongation of the given diagram.

It is, of course, a matter of taste whether one defines strong monoidal functors as oplax monoidal functors with invertible structure maps, as here and in [63], or as lax monoidal functors with invertible structure maps.

Example 3.12 *The underlying functor from $A - $ **mod** *to* $K - $ **v.s.** *for any K-bialgebra is a strict monoidal functor.*

Example 3.13 *When equipped with structure maps induced by inclusions of generators and the universal property of free groups, the free-group functor from* (**Sets**, $\coprod$, ...) *to* (**Grps**, $*$, ...) *becomes a strong monoidal functor.*

And, an example which will be important later:

Example 3.14 *Let A be a K-algebra. Consider the one-object, one-map monoidal category* $\underline{1}$ *with object* $* = I$ *and the obvious monoidal category structure. Then the assignment* $* \mapsto A$ *and* $Id_* \mapsto Id_A$, *with structure maps* $\tilde{F} : F(*) \otimes F(*) \to F(* \otimes *) = m : A \otimes A \to A$ *and* $F_0 : I \to F(I) = 1 : K \to A$ *is a lax monoidal functor from* $\underline{1}$ *to* $K-$**v.s.**

Conversely, every lax monoidal functor from $\underline{1}$ *to* $K - $**v.s.** *is of this form.*

It is an amusing exercise, left to the reader, to verify both of the statements in the last example.

We will be particularly interested in examples of strong monoidal functors whose source is one of our categories of tangles. Theorems which show these to be remarkably easy to construct will occupy much of the balance of this work, though at present we are not in a position to present any examples.

Strong monoidal functors from n-**cobord** to $\mathbb{C}$−**v.s.** or to the category of Hilbert spaces with a suitable monoidal structure turn out to be equivalent to topological quantum field theories as defined by Atiyah [3].

Crucial to the construction of our deformation theories are the coherence theorem of Mac Lane [39] and a non-symmetric variant of the coherence theorem of Epstein [21], which we will soon state in the most convenient form for our purposes.

Definition 3.15 *For any set S, $S \downarrow MonCat$ (resp. $S \downarrow SGCat$) is the category whose objects are (small) monoidal (resp. semigroupal) categories equipped with a map from S to their set of objects, and whose arrows are strict monoidal functors whose object maps commute with the map from S.*

$S \downarrow LaxSGFun$ (resp. $S \downarrow OplaxSGFun$, $S \downarrow StrongSGFun$) is the category whose objects are lax (resp. oplax, strong) semigroupal functors between a pair of semigroupal categories, the source of which is equipped with a map from S to its set of arrows, and whose arrows are pairs of strict monoidal functors forming commuting squares and commuting with the map from S.

Observe that $S \downarrow MonCat$, (resp. $S \downarrow SGCat$, $S \downarrow LaxSGFun$, $S \downarrow OplaxSGFun$ and $S \downarrow StrongSGFun$) is a category of models of an essentially algebraic theory, and thus by general principles has an initial object. We refer to this initial object as the *free monoidal category* (resp. *semigroupal category, lax semigroupal functor, oplax semigroupal functor, strong semigroupal functor*) on S.

Definition 3.16 *A* formal diagram in the theory of monoidal categories (resp. semigroupal categories) *is a diagram in the free monoidal (resp. semigroupal) category on S for some set S.*

A formal diagram in the theory of lax (resp. oplax, strong) semi-groupal functors *is a diagram in the target category of the free lax (resp. oplax, strong) semigroupal functor on S for some set S.*

The coherence theorem of Mac Lane [39] may then be stated as

Theorem 3.17 *Every formal diagram in the theory of monoidal categories commutes. Consequently, any diagram which is the image of a formal diagram under a (strict monoidal) functor commutes.*

Mac Lane's result is the first coherence theorem proven for categories with structure, and its proof is characteristic of proofs of all subsequent coherence theorems. Similar techniques have been used by Epstein [21], Kelly and Laplaza [34], Freyd and Yetter [22], and Shum [48, 49], among others, to prove coherence theorems for other categorical structures. In Chapter 9 we give Shum's result, with proof. As a warm up, we now prove Mac Lane's theorem:

proof: As in all such results, we must begin with a syntactical construction of the free object in question, in this case the free monoidal category $\mathcal{F}(S)$ on a set S.

As objects, we take all non-empty fully parenthesized words on the set S, permitting the inclusion of empty pairs of parentheses. The monoidal product is given by concatenating two objects inside another set of parentheses. The monoidal identity is (). Arrows are named by all formal composites of formal prolongations of the maps of the forms

- $Id_u : u \to u$

- $\alpha_{u,v,w} : ((uv)w) \to (u(vw))$

- $\alpha_{u,v,w}^{-1} : (u(vw)) \to ((uv)w)$

- $\lambda_u : (()u) \to u$

- $\lambda_u^{-1} : u \to (()u)$

- $\rho_u : (u()) \to u$

- $\rho_u^{-1} : u \to (u())$

for u, v, w fully parenthesized words on the set S.

Two formal composites are equivalent when they are equivalent under the equivalence relation which is closed under pre-composition, post-composition and prolongation, and which is generated by all instances of the pentagon, triangles and bigon, all squares which give naturality conditions for the maps names, all triangles which make x and x^{-1} inverses for x any instance of α, λ, or ρ, and all squares which interchange maps to give all instances of the functoriality of the monoidal product involving two prolongations of structure maps.

The monoidal product of two maps $f : x \to y$ and $g : z \to w$ is given by

$$f \otimes g = [f \otimes Id_z][Id_y \otimes g] = [Id_x \otimes g][f \otimes Id_w]$$

where the monoidal products with identity maps are the formal prolongations in our description of maps, and the second equality is one of the squares which provide the functoriality of the monoidal product.

Now, it is immediate by construction that the monoidal category $\mathcal{F}(S)$ just described, together with the inclusion of the set S into the set of objects, is the initial object in $S \downarrow MonCat$. What is not immediately clear is how to provide a more compact description of the arrows in $\mathcal{F}(S)$. Note that $\mathcal{F}(S)$ is a groupoid.

We claim that the connected components of $\mathcal{F}(S)$ are in one-to-one correspondence with words (including the empty word) on the set S, and that between any two objects in the same connected components there is a unique map. Note that proof of these claims will suffice to prove Mac Lane's theorem. To each fully parenthesized word w on S we associate a (possibly empty) $\Upsilon(w)$ word on S by deleting all parentheses. Similarly, to each word v on S we associate a canonical fully parenthesized word: the completely right parenthezised word $R(v)$.

Now for any fully parenthesized word w, construct a canonical map $c_w : w \to R(\Upsilon(w))$ as follows: compose all instances of prolongations of λ and ρ, removing empty parentheses from left to right, and iterate until all empty parentheses are removed; then apply prolongations of α beginning with the outermost applicable instance, and proceeding from left to right if there is more than one outermost instance, until a fully right parenthesized word is obtained as the target. It will be useful to name the initial factor v_w of c_w obtained by composing prolongations of the unit transformations and a second factor γ_w obtained as the composite of prolongations of α. Thus $c_w = \gamma_w(v_w)$.

For any two objects w and v with $\Upsilon(w) = \Upsilon(v)$, the composite $c_w c_v^{-1} : w \to v$ shows that the objects lie in the same connected component.

Conversely, any pair of objects with $\Upsilon(w) \neq \Upsilon(v)$ cannot lie in the same connected component since all of the (prolongations of) generating maps satisfy $\Upsilon(\text{source}(f)) = \Upsilon(\text{target}(f))$.

Now, we must show that there is only one map from any object in a connected component to another. Plainly, this map must be $c_w c_v^{-1}$. In fact, it suffices to show that c_w is the unique map from w to $R(\Upsilon(w))$, since if $\phi : w \to v$ is any map from w to v, then ϕc_v must equal c_w (since it is a map from w to $R(\Upsilon(w))$), and thus c_v equalizes ϕ and $c_w c_v^{-1}$. But maps which are equalized in a groupoid are already equal.

If we consider an arbitrary formal composite ϕ naming a map from w to $R(\Upsilon(w))$, it plainly suffices to show that the triangle $f c_t = c_s$ commutes for $f : s \to t$ any of the formal prolongations of structure maps occuring in ϕ. Moreover, since the category is a groupoid, it suffices to consider only one of each pair of inverse maps.

For $f : w \to v$ a prolongation of λ or ρ, we have $f v_v = v_w$ immediately from the naturality of λ and ρ and the functoriality of $\otimes$.

For prolongations of α, it will be necessary to introduce a syntactic "rank" for parenthesized words which is always reduced in passing from the source to the target of prolongations of λ, ρ or α, since we will want to proceed by induction on the rank.

Let $l(w)$ be the length of w as a word on $S \coprod \{I\}$ when the parentheses are removed, and define $rk(w)$ inductively by

- $rk(s) = 0$ for all $s \in S$

- $rk(I) = 1$

- $rk((vw)) = rk(v) + rk(w) + l(v) + 1$ for v, w parenthesized words on $S \coprod \{I\}$.

Note that applying a prolongation of λ or ρ reduces rk by at least 1, while applying a prolongation of α reduces the rank by the length of the first tensorand of the triple tensor product to which α is applied. It thus follows that the reduced completely right-parenthesized object in each connected component is of minimal rank in the component. Thus, there is nothing to show if the source is of minimal rank (or differs from an object of minimal rank by a single application of a prolongation of a structure map).

For f a prolongation of α we have two distinct cases: that where one (or more) of the three tensorands to which α applies is a parenthesized word of I's (only) and that in which none of them are.

In the first case, it suffices by the naturality of α to consider those instances in which the parenthesized word of I's is simply I. We can then fill the desired triangle $f c_w = c_v$ with three triangles: two of the same type, but with prolongations of λ or ρ in place of f, and a third which is a prolongation of the triangle coherence condition or of one of the triangles of Lemma 3.10.

In the second case, we may assume by the naturality of α and the functoriality of $\otimes$ that the word to which the f applies is reduced (has no instance of I). This case, in turn, reduces to four subcases:

Subcase a) The instance of α in f is the outermost (or left-most of two or more outermost) instances of α applicable to w.

Here there is nothing to do: f is the initial factor of c_w.

Subcase b) The instance of α in f is an outermost, but not the left-most outermost, instance of α.

Here the result follows from the functoriality of $\otimes$.

Subcase c) The instance of α in f applies to a tensorand of the outermost instance of α.

By the naturality of the outermost instance of α, this reduces to an application to a word of lower rank, and thus will follow from the other cases by induction on the rank.

Subcase d) The instance of α in f overlaps with the outermost instance of α so that the two from the initial legs of an instance of the pentagon.

Here we complete the pentagon, and observe that the triangles for all of the arrows of the pentagon other than f follow either directly from another case, or by induction on the rank of the source.

Thus we have proved Mac Lane's coherence theorem. □

The same proof carries the weaker result:

Theorem 3.18 *Every formal diagram in the theory of semigroupal categories commutes. Consequently, any diagram which is the image of a formal diagram under a (strict semigroupal) functor commutes.*

Epstein [21] proves a coherence theorem only for lax semigroupal functors between symmetric semigroupal categories, but the same proof will carry the result:

Theorem 3.19 *Every formal diagram in the theory of lax (resp. oplax, strong) semigroupal functors commutes. Consequently, any diagram which is a functorial image of such a formal diagram under a (strict monoidal) functor commutes.*

These coherence theorems are the basis for a very useful notion and notational convention: throughout our discussion of categorical deformation theory in Part II, we will use *padded composition operators* ⌈ ⌉. These operators are an embodiment of the coherence theorems of Mac Lane [39] and Epstein [21] .

Definition 3.20 *Given a monoidal category $\mathcal{C}$ (resp. a semigroupal functor (whether lax, oplax or strong) $F : \mathcal{X} \to \mathcal{C}$), and a sequence of maps $f_1, \ldots, f_n$ in $\mathcal{C}$ such that the source of f_{i+1} is isomorphic (resp. maps) to the target of f_i by a composition of prolongations of structure* *maps (i.e. by a formal diagram with underlying diagram a chain of composable maps), we let*

$$\lceil f_1, \ldots, f_n \rceil$$

denote the composite $a_0 f_1 a_1 f_2 \ldots a_{n-1} f_n a_n$, where the a_i's are composites of prolongations of structure maps and the following hold:

1. *The source of a_0 is reduced (no tensorands of I) and completely left-parenthesized (resp. reduced and completely left-parenthesized and free from images of monoidal products under F in the lax case, and free from products both of whose factors are images under F in the oplax and strong cases).*

2. *The target of a_n is reduced and completely right-parenthesized (resp. reduced and completely right-parenthesized and free from products both of whose factors are images under F in the lax case, and free from images of monoidal products under F in the oplax and strong cases).*

3. *The composite is well-defined.*

The fact that this defines a well-defined map is a consequence of the coherence theorems.

Observe that $\lceil \ \rceil$ may not be well-defined in the event that there are "accidental coincidences". This may be avoided by replacing the category(ies) with monoidally equivalent categories in which there are no "accidental coincidences", specifically by forming free monoidal categories generated by the given ones, and adjoining a natural isomorphism between the old monoidal product and the new. We will not bother with this here, since in our applications there is another way to remove the potential ambiguity. The reader who is interested may undertake the construction as an exercise after reading Chapter 9, where similar syntactical constructions occur.

In our circumstance, the maps in the sequences to which the padded composition operator is applied will always be components of natural transformations with a particular structure:

Definition 3.21 *Given a monoidal category C (resp. a monoidal functor $F : C \to D$), a natural transformation is C-paracoherent (resp. F-paracoherent) if its source and target functors are iterated prolongations of the structure functors $\otimes$, I, and 1_C (resp. $\otimes$, F, I, 1_C, and 1_D), where I is regarded as a functor from the trivial one object category.*

In the case where the maps in the sequence are specified not merely as maps, but as components of particular paracoherent natural transformations, their sources and targets are given an explicit structure as images of iterated prolongations of structure functors. We may thus require that the "padding" maps given in terms of the structural natural transformations be (components of) natural transformations between the appropriate functors.

Regardless of whether we avoid ambiguity by modifying the category or by restricting the use of $\lceil \ \rceil$ to sequences of paracoherent natural transformations, a number of elementary properties of the operators may be deduced in the case of monoidal categories or strong monoidal functors. In all cases the proofs follow by applying either Mac Lane's or Epstein's coherence theorem. The reader is left to discern what modifications are necessary in the lax and oplax cases.

Lemma 3.22

$$\lceil f_1 \ldots f_n \rceil = \lceil \lceil f_1 \ldots f_k \rceil \lceil f_{k+1} \ldots f_n \rceil \rceil.$$

Lemma 3.23

$$\lceil f_1 \ldots g \otimes I \ldots f_n \rceil = \lceil f_1 \ldots g \ldots f_n \rceil = \lceil f_1 \ldots I \otimes g \ldots f_n \rceil.$$

Lemma 3.24

$$\lceil f_1 \ldots f_n \rceil = \lceil f_1 \ldots \lceil f_k \ldots f_l \rceil \ldots f_n \rceil.$$

Lemma 3.25

$$\lceil f_1 \ldots g \otimes h \ldots f_n \rceil = \lceil f_1 \ldots \lceil g \rceil \otimes h \ldots f_n \rceil = \lceil f_1 \ldots g \otimes \lceil h \rceil \ldots f_n \rceil.$$

Lemma 3.26 *If $\phi_{X_1,\ldots,X_n}$ is a $\mathcal{C}$-paracoherent natural transformation (resp. F-paracoherent natural transformation, for F a strong monoidal functor), then so is $\phi_{X_1,\ldots,I,\ldots,X_n}$, where I is inserted in the ith position, and similarly if I is inserted in the ith position for all $i \in T \subset \{1,\ldots n\}$. Moreover, in this latter case $\lceil \phi_{\ldots} \rceil$ is a paracoherent natural transformation from the fully left-parenthesized product (resp. F of the fully left-parenthesized product) of $X_{i_1} \ldots X_{i_k}$ to the fully right-parenthesized product of $X_{i_1} \ldots X_{i_k}$ (resp. the fully right-parenthesized product of $F(X_{i_1}) \ldots F(X_{i_k})$), where*

$$\{i_1,\ldots,i_k\} = \{1,\ldots,n\} \setminus T$$

and $i_1 < i_2 < \ldots i_n$.

From these lemmas we deduce a final lemma:

Lemma 3.27 *If $\psi_{A,B,C}, \phi_{A,B,C} : [A \otimes B] \otimes C \to A \otimes [B \otimes C]$ are natural transformations, then*

$$\lceil [\phi_{A,I,I} \otimes B] \psi_{A,I,B} \rceil = \lceil \psi_{A,I,B} [\phi_{A,I,I} \otimes B] \rceil$$

and

$$\lceil [A \otimes \phi_{I,I,B}] \psi_{A,I,B} \rceil = \lceil \psi_{A,I,B} [A \otimes \phi_{I,I,B}] \rceil.$$

proof: First, apply Lemma 3.22. Then use the naturality of $\lceil \psi_{A,I,B} \rceil$: $A \otimes B \to A \otimes B$ and the source and target data for $\phi_{A,I,I}$ and $\phi_{I,I,B}$, as given by Lemma 3.26. $\square$

Finally, we make

Definition 3.28 *A monoidal natural transformation is a natural transformation $\phi : F \Rightarrow G$ between monoidal functors which satisfies*

$$\tilde{G}_{A,B}(\phi_{A \otimes B}) = \phi_A \otimes \phi_B(\tilde{F}_{A,B})$$

and $F_0 = G_0(\phi_I)$. A semigroupal natural transformation between semigroupal functors is defined similarly.

and

Definition 3.29 *A* monoidal equivalence *between monoidal categories* C *and* D *is an equivalence of categories in which the functors* $F : C \to D$ *and* $G : D \to C$ *are equipped with the structure of monoidal functors, and the natural isomorphisms* $\phi : FG \Rightarrow Id_C$ *and* $\psi : GF \Rightarrow Id_D$ *are both monoidal natural transformations. If there exists a monoidal equivalence between* C *and* D, *we say that* C *and* D *are* monoidally equivalent.

$$F([A \otimes B] \otimes C) \xrightarrow{\quad F(\alpha) \quad} F(A \otimes [B \otimes C])$$

$F_{A \otimes B, C}$

$F_{A, B \otimes C}$

$$F(A \otimes B) \otimes F(C) \qquad\qquad F(A) \otimes F(B \otimes C)$$

$F_{A,B} \otimes F(C)$

$F(A) \otimes F_{B,C}$

$$[F(A) \otimes F(B)] \otimes F(C) \xrightarrow{\quad\quad} F(A) \otimes [F(B) \otimes F(C)]$$
$$\alpha$$

$$F(I \otimes A) \xrightarrow{\quad F(\lambda) \quad} F(A)$$

$F_{I,A}$

λ

$$F(I) \otimes F(A) \xleftarrow{\quad\quad} I \otimes F(A)$$
$$F_0 \otimes F(A)$$

$$F(A \otimes I) \xrightarrow{\quad F(\rho) \quad} F(A)$$

$F_{A,I}$

ρ

$$F(A) \otimes F(I) \xleftarrow{\quad\quad} F(A) \otimes I$$
$$F(A) \otimes F_0$$

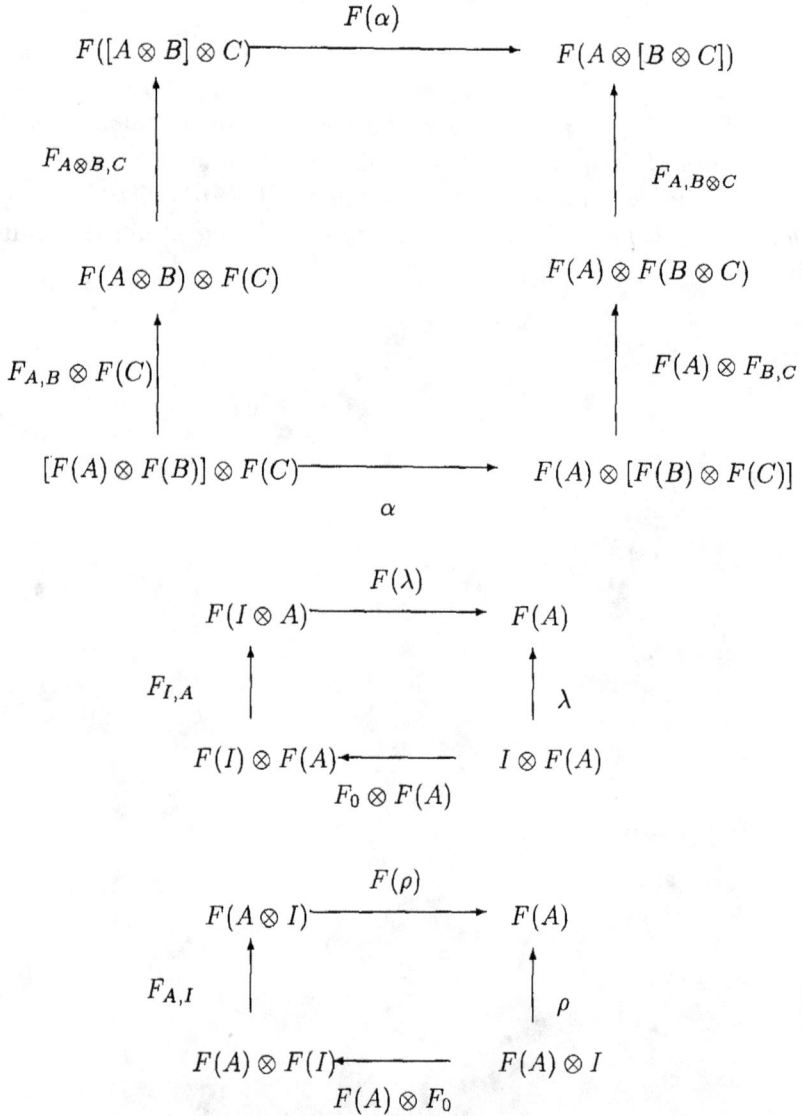

Figure 3.2: Coherence Conditions for a Lax Monoidal Functor

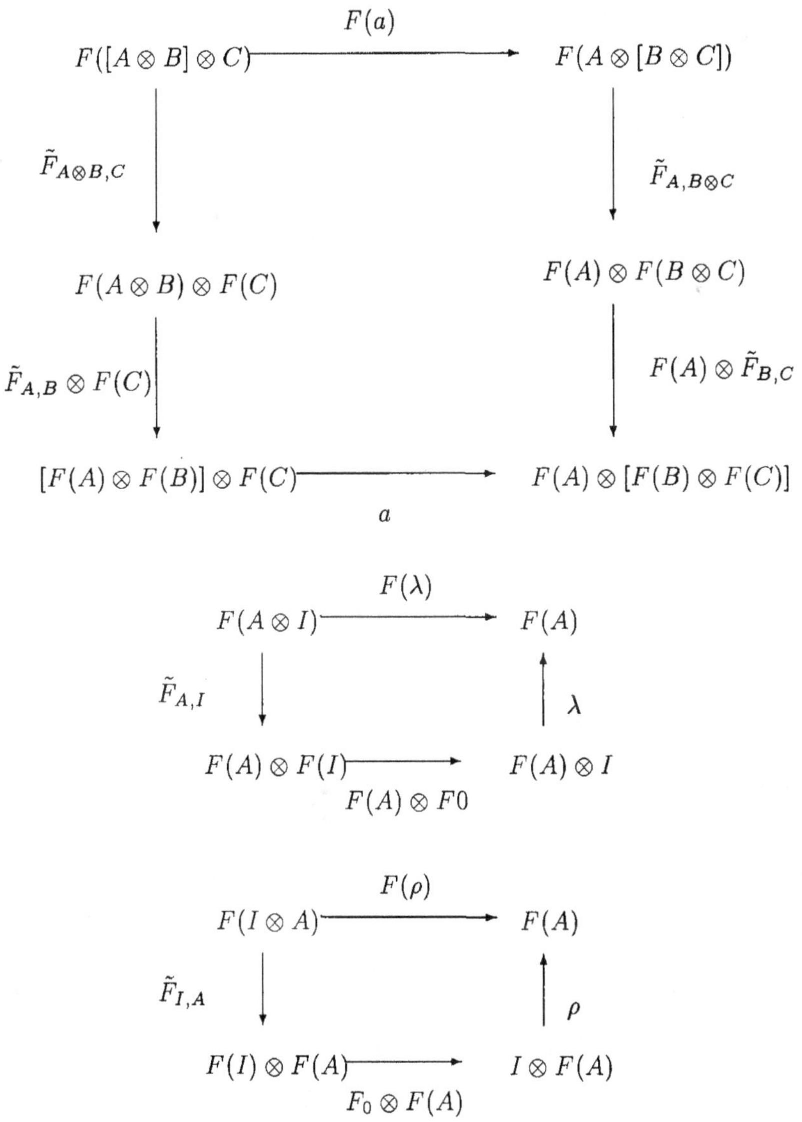

$$F([A \otimes B] \otimes C) \xrightarrow{\quad F(a) \quad} F(A \otimes [B \otimes C])$$

$$\tilde{F}_{A \otimes B, C} \downarrow \qquad\qquad\qquad \downarrow \tilde{F}_{A, B \otimes C}$$

$$F(A \otimes B) \otimes F(C) \qquad\qquad F(A) \otimes F(B \otimes C)$$

$$\tilde{F}_{A,B} \otimes F(C) \downarrow \qquad\qquad\qquad \downarrow F(A) \otimes \tilde{F}_{B,C}$$

$$[F(A) \otimes F(B)] \otimes F(C) \xrightarrow{\quad a \quad} F(A) \otimes [F(B) \otimes F(C)]$$

$$F(A \otimes I) \xrightarrow{\quad F(\lambda) \quad} F(A)$$

$$\tilde{F}_{A,I} \downarrow \qquad\qquad \uparrow \lambda$$

$$F(A) \otimes F(I) \xrightarrow{\quad F(A) \otimes F0 \quad} F(A) \otimes I$$

$$F(I \otimes A) \xrightarrow{\quad F(\rho) \quad} F(A)$$

$$\tilde{F}_{I,A} \downarrow \qquad\qquad \uparrow \rho$$

$$F(I) \otimes F(A) \xrightarrow{\quad F_0 \otimes F(A) \quad} I \otimes F(A)$$

Figure 3.3: Coherence Conditions for an Oplax Monoidal Functor

Chapter 4

A Digression on Algebras

In order to introduce promptly the important Examples 3.7 and 3.12, we were obliged to make a definition of bialgebras (Definition 3.6) which was unusual, both in that it included a category theoretic nicety usually omitted by virtue of Theorem 3.17, and in that it does not invoke the notion of coalgebra.

In this chapter, we recall the usual approach to algebras, coalgebras, bialgebras and Hopf algebras. We will now suppress the writing of structure maps for the monoidal structure on $K - $ **v.s.** by Theorem 3.17.

Definition 4.1 *A* (unital) K-*algebra, for* K *a field, is a vector space* A *over* K *equipped with linear maps* $m : A \otimes_K A \to A$, *called the* multiplication *, and* $1 : K \to A$, *called the* unit *and satisfying*

$$A \otimes A \otimes A \xrightarrow{\quad m \otimes A \quad} A \otimes A$$

$$A \otimes m \Big\downarrow \qquad\qquad \Big\downarrow m$$

$$A \otimes A \xrightarrow{\quad m \quad} A$$

and

$$A \otimes K \xleftarrow{\quad \rho^{-1} \quad} A \xrightarrow{\quad \lambda^{-1} \quad} K \otimes A$$

$$A \otimes 1 \Big\downarrow \qquad Id_A \Big\downarrow \qquad\qquad \Big\downarrow 1 \otimes A$$

$$A \otimes A \xrightarrow{\quad m \quad} A \xleftarrow{\quad m \quad} A \otimes A$$

Observe that by virtue of the universal property of $\otimes$, this is equivalent to the elementary definition of algebra as a K-vectorspace with a unital ring structure whose addition coincides with the vectorspace addition, and whose multiplication is K-bilinear (cf. e.g. [4]). Any reader unfamiliar with the definition just given, who was consequently puzzled by the suggested exercise after Example 3.14, should certainly now undertake the exercise.

This definition, on the other hand, has the virtue that it generalizes nicely to monoidal categories other than $K - \mathbf{v.s.}$ (the generalized notion is usually called a *monoid* in the monoidal category, using the name in the case of (**Sets**, $\times$, ...)). Consequently, an analogous definition can be made in $K - \mathbf{v.s.}^{op}$, but then interpreted as a definition in $K - \mathbf{v.s.}$ "dual to" the definition of algebra:

Definition 4.2 *A* (counital) *K-coalgebra, over a field K, is a vectorspace A over K equipped with linear maps $\Delta : A \to A \otimes_K A$, called the* comultiplication, *and $\epsilon : A \to K$, called the* counit, *and satisfying*

$$
\begin{array}{ccc}
A \otimes A \otimes A & \xleftarrow{A \otimes \Delta} & A \otimes A \\
\big\uparrow{\scriptstyle \Delta \otimes A} & & \big\uparrow{\scriptstyle \Delta} \\
A \otimes A & \xleftarrow{\Delta} & A
\end{array}
$$

and

$$
\begin{array}{ccccc}
A \otimes K & \xrightarrow{\rho} & A & \xleftarrow{\lambda} & K \otimes A \\
\big\uparrow{\scriptstyle A \otimes \epsilon} & & \big\uparrow{\scriptstyle Id_A} & & \big\uparrow{\scriptstyle \epsilon \otimes A} \\
A \otimes A & \xleftarrow{\Delta} & A & \xrightarrow{\Delta} & A \otimes A
\end{array}
$$

As is the case with the more familiar algebras, coalgebras are quite plentiful:

Example 4.3 *Consider any vector space V with a specified basis B. V has a coalgebra structure given on the basis elements by $\Delta(b) = b \otimes b$ and $\epsilon(b) = 1$.*

Elements of any coalgebra on which Δ and ϵ are given by the formulas of the previous example are called *grouplike* elements for reasons which will become clear in connection with our examples of bialgebras given below.

A very general construction of coalgebras exists, which arises often in their application to combinatorial and topological problems: suppose we are given a category with coproducts and an initial object, 0, such that every object admits at most finitely many expressions as a coproduct of two other objects. Then for any collection of objects, or more properly isomorphism classes, $\mathcal{A}$, which is "closed under summand", that is such that $A = A_1 \coprod A_2$ and $A \in \mathcal{A}$ implies $A_1, A_2 \in \mathcal{A}$, the vectorspace $V_{\mathcal{A}}$ with basis $\mathcal{A}$ has a natural coalgebra structure given by

$$\Delta(A) = \sum_{A = A_1 \coprod A_2} A_1 \otimes A_2$$

and

$$\epsilon(A) = \begin{cases} 1 & \text{if } A = 0 \\ 0 & \text{if } A \neq 0. \end{cases}$$

Example 4.4 *Let Γ denote the collection of isomorphism classes of finite graphs, including the empty graph. Then V_{Γ} has a coalgebra structure given by the construction above.*

Similar examples can be constructed using operations with properties similar to coproducts, as, for example:

Example 4.5 *Let $\mathcal{L}$ denote the collection of all isotopy classes of tame links in $\mathbb{R}^3$ including the empty link $\emptyset$. Then $V_{\mathcal{L}}$ has a coalgebra structure given by*

$$\Delta(A) = \sum_{L=L_1+L_2} L_1 \otimes L_2$$

and

$$\epsilon(A) = \begin{cases} 1 & \text{if } A = \emptyset \\ 0 & \text{if } A \neq \emptyset \end{cases}$$

where $+$ denotes separated union of links.

Definition 4.6 *A K-algebra homomorphism (resp. K-coalgebra homomorphism) from one algebra (resp. coalgebra) A to another B is a K-linear map $h : A \to B$ such that*

$$m_B(h \otimes h) = h(m_A) \text{ and } h(1_A) = 1_B$$

(resp.

$$[h \otimes h](\Delta_A) = \Delta_B(h) \text{ and } \epsilon_A = \epsilon_B(h) \text{)}$$

hold.

We can now replace Definition 3.6 with a definition that makes the symmetries of the notion of bialgebra clear:

Definition 4.7 *A bialgebra over a field K is a K-vectorspace equipped with a K-algebra structure $m : A \otimes A \to A$, $1 : K \to A$, and a K-coalgebra structure $\Delta : A \to A \otimes A$, $\epsilon : A \to K$, and satisfying any (and thus all) of the following equivalent conditions:*

1. Δ and ϵ are K-algebra homomorphisms

2. *m and* 1 *are* K*-coalgebra homomorphisms (K has an obvious (trivial) coalgebra structure, while a coalgebra structure on* $A \otimes A$ *can be found by dualizing the construction of the algebra structure on* $A \otimes A$ *mentioned in Definition 3.6.).*

Thus, it can be seen that the conditions defining a bialgebra are self-dual. They reduce, in fact, to two self-dual diagrams relating the unit and counit, a pair of dual diagrams relating the unit and comultiplication and the counit and multiplication, and the diagram of Figure 4.1. Writing out the other three diagrams is left as an (easy) exercise for the reader.

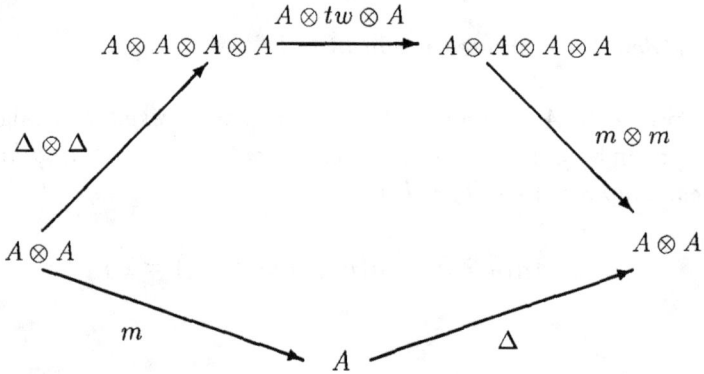

Figure 4.1: The Relationships of the Multiplication and Comultiplication in a Bialgebra

A homomorphism of bialgebras is then defined in the obvious way.

Example 4.8 *Let* G *be a group. Then the group algebra* $K[G]$ *is a bialgebra when equipped with the comultiplication and counit given on the basis of elements of* G *by*

$$\Delta(g) = g \otimes g$$

$$\epsilon(g) = 1.$$

The coproduct and counit formulae appropriate to the bialgebra structure on group algebras is the reason for the name "grouplike" applied to those elements of any coalgebra which have coproduct and counit given by the same prescription.

Example 4.9 *Let $\mathfrak{g}$ be a Lie algebra, and let $U(\mathfrak{g})$ be its universal enveloping algebra. Then $U(\mathfrak{g})$ is a bialgebra when equipped with the comultiplication and counit given on the elements of $\mathfrak{g}$ (which generate $U(\mathfrak{g})$ as a unital algebra) by*

$$\Delta(\gamma) = \gamma \otimes 1 + 1 \otimes \gamma$$

$$\epsilon(\gamma) = 0.$$

A rather important class of examples can be derived from those in Example 4.9 by algebraic deformation in the manner of Gerstenhaber and Schack (cf. [23, 24, 25]), although they arose historically as deformations in a less precise sense. These are, of course, the so-called "quantized universal enveloping algebras" (QUEAs), which were discovered independently by Drinfel'd [16] and Jimbo [28].

More precisely, following Reshetikhin [45]:

Definition 4.10 *The QUEA $U_q(\mathfrak{g})$ is the algebra with generators $X_i^{\pm}$ and H_i for $i = 1, \ldots, r$, where r is the rank of $\mathfrak{g}$, and with relations:*

$$[H_i, H_j] = 0, \quad [H_j, X_i^{\pm}] = \pm(\alpha_j, \alpha_i)X_i^{\pm}$$

$$[X_i^+, X_j^-] = \delta_{ij}\frac{\sinh(\frac{h}{2}H_i)}{\sinh(\frac{h}{2})}$$

and for $i \neq j$

$$\sum_{k=0}^{n}(-1)^k \begin{pmatrix} n \\ k \end{pmatrix}_{q_i} q_i^{-\frac{k(n-k)}{2}}(X_i^{\pm})^k X_j^{\pm}(X_i^{\pm})^{n-k} = 0,$$

where $q = e^h$, $q_i = q^{\frac{(\alpha_i, \alpha_i)}{2}}$, and the α_i are roots, with $(-, -)$ denoting the usual scalar product on the roots satisfying $A_{ij} = (\alpha_i, \alpha_j)(\alpha_j, \alpha_j)$ for $[A_{ij}]$ the Cartan matrix of $\mathfrak{g}$.

The algebra thus defined admits a bialgebra structure given by

$$\Delta(H_i) = H_i \otimes 1 + 1 \otimes H_i$$

$$\Delta(X_i^{\pm}) = X_i^{\pm} \otimes q^{\frac{H_i}{4}} + q^{-\frac{H_i}{4}} \otimes X_i^{\pm}$$

and

$$\epsilon(H_i) = 0, \quad \epsilon(X_i^{\pm}) = 0.$$

The construction of coalgebras given above can be used to construct bialgebras by applying it to the algebra freely generated by the collection of combinatorial objects. For example, $K[\Gamma]$ and $K[\mathcal{L}]$ each have bialgebra structures given on generators by the coalgebra structures of Examples 4.4 and 4.5 respectively.

Several additional structures on a bialgebra will correspond to additional useful structures on the monoidal category A-**mod**:

Definition 4.11 *A* Hopf *algebra H is a bialgebra equipped with a linear map $S : H \to S$ called the* antipode, *which satisfies the relations of Figure 4.2.*

The bialgebras of Examples 4.8, 4.9, and 4.10 are Hopf algebras. In the first, the antipode is given on the elements of G by $S(g) = g^{-1}$. In the second, the antipode is given on elements of $\mathfrak{g}$ by $S(\gamma) = -\gamma$. In the third, the antipode is given by $S(H_i) = -H_i$ and $S(X_i^{\pm}) = -q^r X_i^{\pm} q^r$, where $r = \frac{1}{4} \sum_{\alpha \in \Delta_+} H_\alpha$, for Δ_+ the positive roots of $\mathfrak{g}$ and H_α given by $H_\alpha = \sum_{i=1}^n H_i$ whenever $\alpha = \sum_{i=1}^n \alpha_i$.

Also, following Drinfel'd [17] we make

Definition 4.12 *A bialgebra A is* quasi-triangular *(resp.* triangular*) if it is equipped with a unit $R \in A \otimes A$ satisfying*

$$tw(\Delta(a)) = R\Delta(a)R^{-1}$$

$$(\Delta \otimes Id_A)(R) = R_{13}R_{23}$$

and

$$(Id_A \otimes \Delta)(R) = R_{13}R_{12}$$

(resp. such a unit with $Rtw_{A,A}(R) = 1 \otimes 1$).

Here, $R_{12} = R \otimes 1$, $R_{23} = 1 \otimes R$, and for $R = \sum_i a_i \otimes b_i$, $R_{13} = \sum_i a_i \otimes 1 \otimes b_i$.

The bialgebras of Examples 4.8 and 4.9 are triangular Hopf algebras with $R = 1 \otimes 1$, while the bialgebras of Example 4.10 are quasi-triangular Hopf algebras. For the present, the exact definition of R for the Hopf algebras of Example 4.10 does not concern us. The reader is referred to Reshetikhin [45] for a construction.

In fact, the Hopf algebras of Example 4.10 satisfy an additional condition:

Definition 4.13 *A* ribbon Hopf algebra *is a quasi-triangular Hopf algebra (A, R) equipped with a central unit v satisfying*

$$v^2 = uS(u), \quad S(v) = v, \quad \epsilon(v) = 1,$$

and

$$\Delta(v) = (tw_{A,A}(R)R)^{-1}(v \otimes v)$$

where $u = \cdot([S \otimes Id_A](tw_{A,A}(R)))$.

Each of these definitions may seem a little arcane until it is observed that each corresponds to a perfectly reasonable structure on the category of modules over A with the monoidal structure induced by $\otimes_K$. We will introduce the relevant structures in the next chapter.

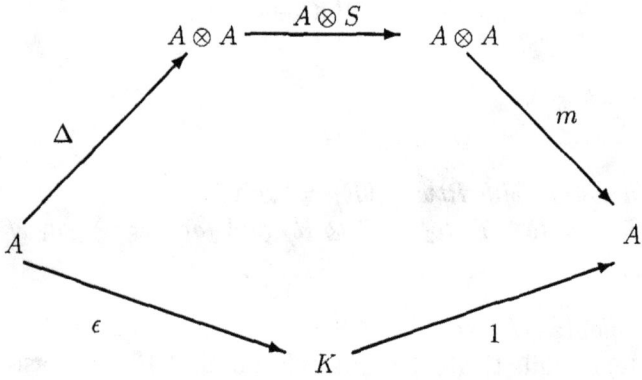

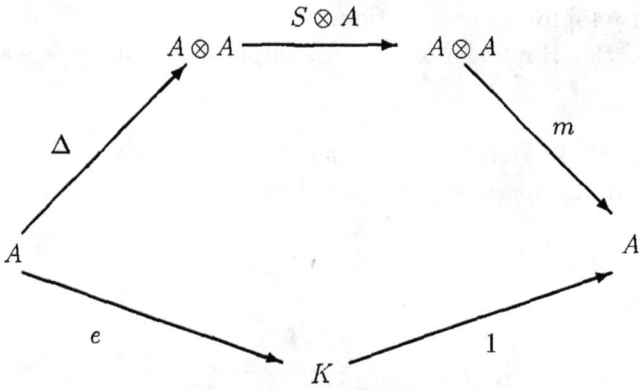

Figure 4.2: Defining Conditions for an Antipode

Chapter 5

More About Monoidal Categories

We will be concerned with monoidal categories with additional structure.

Definition 5.1 *A* braided monoidal category *is a monoidal category equipped with a monoidal natural isomorphism* $\sigma : \otimes \Rightarrow \otimes(tw)$, *called the* braiding *, where* $tw : \mathcal{C} \times \mathcal{C} \to \mathcal{C} \times \mathcal{C}$ *is the "twist functor"* $(tw(f, g) = (g, f))$ *and* σ *satisfies the two relations of Figure 5.1. In Figure 5.1 the sign indicating inversion or non-inversion of* σ *must be chosen consistently throughout.*

A braided monoidal category *is a* symmetric monoidal category *if the components of* σ *satisfy* $\sigma_{B,A}(\sigma_{A,B}) = 1_{A \otimes B}$ *for all objects A and B, in which case the braiding is called the* symmetry *.*

Examples 3.2, 3.3, 3.4 and 3.5 are all symmetric monoidal categories, with fairly obvious structure maps, while Example 3.7 may be symmetric, braided, or neither, depending on whether the bialgebra A is triangular, quasi-triangular, or neither.

The following example explains the name "braided" monoidal category:

71

$$
\begin{array}{ccc}
A \otimes (B \otimes C) & \xrightarrow{\;\sigma^{\pm 1}\;} & (B \otimes C) \otimes A
\end{array}
$$

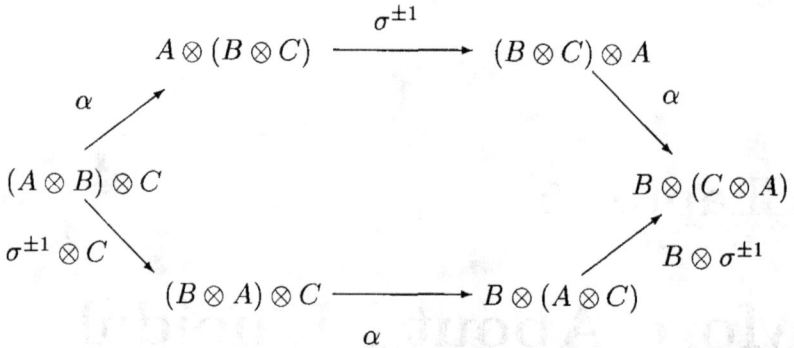

Figure 5.1: The Hexagon

Definition 5.2 *The n-strand Artin braid group, B_n, is the group*

$$
\pi_1([\mathbb{C}^n \setminus \Delta]/\mathfrak{S}_n)
$$

where Δ is the "big diagonal", that is

$$
\Delta = \{(z_1, \ldots, z_n) | \exists i \neq j \ s.t. \ z_i = z_j\},
$$

and the symmetric group $\mathfrak{S}_n$ acts by permuting the coördinates.

Observe that the elements of $[\mathbb{C}^n \setminus \Delta]/\mathfrak{S}_n$ may be identified with n-element subsets of $\mathbb{C}$. Also note that B_0 and B_1 are both trivial groups.

The identification of the elements of $[\mathbb{C}^n \setminus \Delta]/\mathfrak{S}_n$ as subsets of the plane presents a nice geometric interpretation of elements in B_n that accounts for the name and connects the braid groups with our category of tangles:

Fix a base-point, say the subset $B = \{\frac{1}{n+1} + \frac{i}{2}, \ldots, \frac{n}{n+1} + \frac{i}{2}\}$, and consider a representative p of an element $[p]$ in B_n (with this base-point). For each $t \in \mathbb{I}$, $p(t)$ is a subset of the plane. We may then "graph" these subsets in $\mathbb{C} \times \mathbb{I}$, giving rise to a family of n disjoint arcs intersecting each level $\mathbb{C} \times \{t\}$ in an n-element set. Observe that a homotopy in

$[\mathbb{C}^n \setminus \Delta]/\mathfrak{S}_n$ corresponds to a level-preserving isotopy of the family of curves (as a 1-submanifold with boundary). It is an easy technical lemma to show that such a level-preserving isotopy is a restriction of an ambient isotopy rel boundary, which is, moreover, trivial outside $R \times \mathbb{I}$ for some rectangle R.

The 1-submanifold constructed in this way is called a *geometric braid*. We have almost shown that geometric braids are a special case of tangles. All that remains is to observe that there exists a homotopy in $[\mathbb{C}^n \setminus \Delta]/\mathfrak{S}_n$ (and consequently an ambient isotopy rel boundary in $\mathbb{C} \times \mathbb{I}$) which contracts the entire space into the cube $[\mathbb{I} + i\mathbb{I}] \times \mathbb{I}$, while fixing a neighborhood of $B \times \mathbb{I}$.

The equivalence relation on braids is *a priori* weaker than that on tangles. However, we have:

Theorem 5.3 (Artin [2]) *If two geometric braids are ambient isotopic rel boundary, then they are ambient isotopic by a level-preserving isotopy.*

Theorem 5.4 (Artin [2]) *The group B_n admits a presentation by generators and relations of the following form:*

$$\langle \sigma_1, \ldots \sigma_{n-1} \mid \sigma_i \sigma_{i+1} \sigma_i = \sigma_{i+1} \sigma_i \sigma_{i+1} \text{ for } i = 1, \ldots n - 2;$$
$$\sigma_i \sigma_j = \sigma_j \sigma_i \text{ for } |i - j| > 1 \rangle.$$

Now, in terms of this presentation it is easy to see that there is a group homomorphism $\phi_{n,m} : B_n \times B_m \to B_{n+m}$ given on generators by

$$\phi_{n,m}(\sigma_i, 1) = \sigma_i \quad \phi_{n,m}(1, \sigma_j) = \sigma_{j+n}.$$

Observe that this is enough, since

1. $B_n \times B_m$ is generated by the elements of the forms $(\sigma_i, 1)$ and $(1, \sigma_j)$,

2. all relations involving generators of only one of the two forms are given by the relations from B_n (resp. B_m) on the first (resp. second) coördinates of elements of the first (resp. second) form,

3. a complete set of relations for a product is given on generators of the forms $(x, 1)$ and $(1, y)$ by relations from the two groups as in the previous item and all relations of the form $(x, 1)(1, y) = (1, y)(x, 1)$,

4. the map described above preserves all of the relations of the previous two items.

Example 5.5 *The category* **Braids** *given by*

$$Ob(\mathbf{Braids}) = \mathbb{N}$$

with maps given by

$$\mathbf{Braids}(n, m) = \left\{ \begin{array}{ll} \emptyset & \textit{if } n \neq m \\ B_n & \textit{if } n = m \end{array} \right.$$

with composition given by the group laws in each braid group, admits a strict monoidal structure given on objects by $n \otimes m = n + m$ and on maps by $\beta \otimes \gamma = \phi_{n,m}(\beta, \gamma)$, where $\beta : n \to n$ and $\gamma : m \to m$, with $I = 0$. Moreover, this monoidal structure is equipped with a braiding given in terms of geometric braids by the maps $\sigma_{n,m}$ which pass the first n strands of $n+m$ behind the last m. (Equivalently, it suffices to specify that $\sigma_{1,1} = \sigma_1 \in \mathbf{Braids}(2, 2) = B_2$.)

The following theorem of Joyal and Street [30] then fully justifies the name:

Theorem 5.6 *The braided monoidal category freely generated by a single object is monoidally equivalent to* **Braids**.

proof: We have already observed above that **Braids** is a braided monoidal category. Now, every object of **Braids** is a monoidal product of copies of 1 (taking 0 as the empty monoidal product); the monoidal structure is strict; and it is clear that all maps are compositions of prolongations of σ. Thus **Braids** is generated as a braided monoidal category by 1, and there is a unique strict monoidal functor $B : \mathbf{F}(1) \to$ **Braids**, where $\mathbf{F}(1)$ is the free braided monoidal category generated by 1.

Thus we wish to show that there is a strong monoidal functor

$$S : \mathbf{Braids} \to \mathbf{F}(1)$$

and natural isomorphisms $\eta : BS \Rightarrow 1_{\mathbf{F}(1)}$ and $\epsilon : SB \Rightarrow 1_{\mathbf{Braids}}$.

At the level of objects, S is given by mapping n to the fully left-parenthesized monoidal product of n 1's in $\mathbf{F}(1)$. On maps, S is given by mapping an n-strand braid $\sigma_{i_1} \ldots \sigma_{i_m}$ to a composite of the form

$$\lceil s_1 s_2 \ldots s_m \rceil.$$

where s_j is a left-parenthesized monoid product of $i_j - 1$ copies of Id_1, a copy of $\sigma_{1,1}$ and $n - i_j - 1$ more copies of Id_1.

It is immediate by construction that $SB = 1_{\mathbf{Braids}}$, so that ϵ may be taken to be the identity natural transformation. It is easy to verify that the structural natural transformations for S as a monoidal functor and the components of η are given by coherence maps from Mac Lane's coherence theorem. $\square$

One thing which can be immediately observed is that every braided monoidal category admits a second braided monoidal structure with the same underlying monoidal category, namely, the one obtained by taking $[\sigma_{B,A}]^{-1} : A \otimes B \to B \otimes A$. From this point of view, symmetric monoidal categories are those for which the two monoidal structures coincide.

Of course, as often has been observed, in category theory one cannot expect or demand coincidence, only isomorphism. Thus it is reasonable to consider braided monoidal categories equipped with an isomorphism in a suitable sense between these two braided monoidal structures. Thus we make:

Definition 5.7 *A braided monoidal category* $(\mathcal{V}, \otimes, I, \alpha, \rho, \lambda, \sigma)$ *is balanced if it is equipped with a natural automorphism* $\theta : 1_{\mathcal{V}} \Rightarrow 1_{\mathcal{V}}$ *called the* balancing *or* twist map *satisfying*

$$\theta_I = Id_I$$

and

$$\theta_{A \otimes B} = \sigma_{B,A}(\sigma_{A,B}(\theta_A \otimes \theta_B)).$$

Notice that the second condition may be rewritten as

$$[\sigma_{B,A}]^{-1}(\theta_{A \otimes B}) = \sigma_{A,B}(\theta_A \otimes \theta_B).$$

In Chapter 12 we will give another characterization of braided monoidal categories in terms of a "multiplication" on a monoidal category.

Another concept familiar from the case of categories of vector-spaces is the notion of a dual object.

Definition 5.8 *A* right *(resp.* left*) dual to an object* X *in a monoidal category* $\mathcal{V} = (\mathcal{V}, \otimes, I, \alpha, \lambda, \rho)$ *is an object* X^* *(resp.* $^*X$*) equipped with maps* $\epsilon : X \otimes X^* \to I$ *and* $\eta : I \to X^* \otimes X$ *(resp.* $e :^* X \otimes X \to I$ *and* $h : I \to X \otimes\, ^*X$*) such that the composites*

$$X \xrightarrow{\rho^{-1}} X \otimes I \xrightarrow{X \otimes \eta} X \otimes (X^* \otimes X) \xrightarrow{\alpha^{-1}} (X \otimes X^*) \otimes X \xrightarrow{\epsilon \otimes X} I \times X \xrightarrow{\lambda} X$$

and

$$X^* \xrightarrow{\lambda^{-1}} I \otimes X^* \xrightarrow{\eta \otimes X^*} (X^* \otimes X) \otimes X^* \xrightarrow{\alpha} X^* \otimes (X \otimes X^*) \xrightarrow{X^* \otimes \eta} X^* \otimes I \xrightarrow{\rho} X^*$$

(resp.

$$X \xrightarrow{\lambda^{-1}} I \otimes X \xrightarrow{h \otimes X} (X \otimes\, ^*X) \otimes X \xrightarrow{\alpha} X \otimes (^*X \otimes X) \xrightarrow{X \otimes e} X \otimes I \xrightarrow{\rho} X$$

and

$$^*X \xrightarrow{\rho^{-1}} {}^*X \otimes I \xrightarrow{{}^*X \otimes h} {}^*X \otimes (X \otimes\, ^*X) \xrightarrow{\alpha^{-1}} (^*X \otimes X) \otimes\, ^*X \xrightarrow{e \otimes\, ^*X} I \otimes\, ^*X \xrightarrow{\lambda^*} X)$$

are identity maps.

Notice that in the case of a symmetric monoidal category

$$(\mathcal{V}, \otimes, I, \alpha, \rho, \lambda, \sigma),$$

a right dual to any object is canonically a left dual by taking $e = \sigma \epsilon$ and $h = \eta \sigma$.

This type of duality is an abstraction from the sort of duality which exists in categories of finite dimensional vector-spaces. It is not hard to show that the canonical isomorphism from the second dual of a vector-space to the space generalizes to give canonical isomorphisms $k :^* (X^*) \to X$ and $\kappa : (^*X)^* \to X$. In general, however, there may not even be any maps from X^{**} or $^{**}X$ to X (cf. [22]). In cases where every object admits a right (resp. left) dual, it is easy to show that a choice of right (resp. left) dual for every object extends to a contravariant functor, whose application to maps will be denoted f^* (resp. $^*f$), and that the canonical maps noted above become natural isomorphisms between the compositions of these functors and the identity functor. Likewise, it is easy to show that $(A \otimes B)^*$ is canonically isomorphic to $B^* \otimes A^*$, and similarly for left duals.

In the case of a braided monoidal category every right dual is also a left dual, but in general the left dual structure is non-canonical (cf. [22]). In symmetric monoidal categories, we return to the familiar: right duals are canonically left duals. In non-symmetric braided monoidal categories it is possible to provide a canonical left dual structure on all right duals only in the presence of additional structure on the category: the category must be balanced and the balancing be related to the duality structure in a natural way.

Definition 5.9 *A braided monoidal category C is* ribbon (or tortile) *if all objects admit right duals, and it is equipped with a balancing $\theta : 1_C \Rightarrow 1_C$, which moreover satisfies*

$$\theta_{A^*} = \theta_A^*.$$

Definition 5.10 *A symmetric monoidal category C is* rigid *if all objects admit (right) duals.*

$$X^{**} \xrightarrow{\quad \phi_{X^*} \quad} {}^*(X^*)$$

$$(\phi_X^{-1})^* \downarrow \qquad\qquad\qquad \downarrow k$$

$$({}^*X)^* \xrightarrow[\quad \kappa \quad]{} X$$

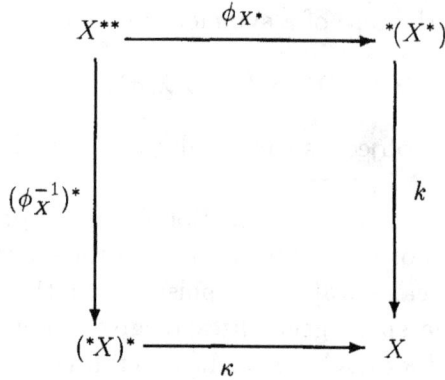

Figure 5.2: Two-Sided Dual Condition

Definition 5.11 *A monoidal category is* sovereign *if it is equipped with a choice for each object X of a right dual X^* and a left dual ${}^*X$, and a natural isomorphism $\phi_X : X^* \to {}^*X$ satisfying the condition of Figure 5.2.*

We then have the following theorem, which is due to Deligne [15] (cf. also [61], where a more detailed proof may be found):

Theorem 5.12 *Every ribbon category is a sovereign category when equipped with the left-duals obtained by letting ${}^*X = X^*$ with structure given by $e = \sigma^{-1}\epsilon$ and $h = \eta\sigma$, and conversely.*

sketch of proof: The proof is reduced to a sequence of lemmas. Throughout, we use Mac Lane's coherence theorem to justify the suppression of all instances of monoidal structure maps.

Lemma 5.13 *The identity maps on the right duals are components of a natural isomorphism from the right dual functor to the left dual functor (with the given structure maps).*

sketch of proof of Lemma 5.13: This amounts to saying that the functors ${}^*(-)$ and $(-)^*$ are equal. This is immediate by construction for objects, but must be checked for maps. The reader familiar with the diagrams that can be used to represent maps in braided monoidal categories can easily recover the proof given in [61]. Briefly, one first shows that

$$^*f = [^*X \otimes h_X][e_X \otimes X^*]f^*$$

by using the naturality of σ (twice) and the right duality structure of $(-)^*$ (once). One then uses the left duality structure to obtain the desired result. $\square$

Lemma 5.14 *Any natural isomorphism* $\phi : X^* \to {}^*X$ *is induced by a natural automorphism of the identity functor* $\theta : X \to X$, *and conversely any natural automorphism of the identity functor induces a natural isomorphism from* $(-)^*$ *to* $^*(-)$.

proof of Lemma 5.14 This is immediate from the previous lemma and the dinaturality properties of ϵ and η. Given ϕ, θ is given by

$$\theta_X = [\eta_X \otimes X][\phi_X \otimes X \otimes X]\sigma_{X,X}[e \otimes X],$$

while given θ, ϕ is given by

$$\phi_X = [h \otimes X^*]\sigma^{-1}_{^*X,X^*}[\theta_X \otimes X^* \otimes {}^*X][\epsilon_X \otimes {}^*X].$$

$\square$

Lemma 5.15 *A natural isomorphism* $\phi : X^* \to {}^*X$ *provides a sovereign category structure for the right and left dual structures given in the statement of the theorem if and only if the corresponding natural automorphism* $\theta : Id_C \Rightarrow Id_C$ *satisfies the balancing axioms of Definition 5.7.*

sketch of proof of Lemma 5.15

The proof that the balancing condition implies sovereignty is done by calculating the two composites in the diagram obtained from that of Figure 5.2 by inverting both vertical maps. By using the naturality conditions on the braiding and the dinaturality of the structure maps for the right duals, it follows that $\kappa_X k_X^{-1}$ equals

$$[h_{\bullet X} \otimes (^*X)^*][^*X \otimes \sigma_{(X^*)^\bullet,(^\bullet X)^\bullet}][\epsilon_{\bullet X} \otimes (X^*)^*].$$

(Recall that for any object Y, $^*Y = Y^*$.) Observing that $\theta_I = Id_I$, it follows from the naturality of θ that

$$\kappa_X k_X^{-1} =$$
$$[h_{\bullet X} \otimes (^*X)^*][^*X \otimes \sigma_{(X^*)^\bullet,(^\bullet X)^\bullet}][\theta_{\bullet X \otimes (^\bullet X)^\bullet} \otimes {}^*(X^*)][\epsilon_{\bullet X} \otimes {}^*(X^*)].$$

Similarly, recalling the definition of ϕ in terms of θ and the definition of $(-)^*$ on maps, one can use the triangle condition and dinaturality of the unit and counit of the structure maps for $(-)^*$ and the naturality and invertibility of the braiding to show that

$$\phi_X^* \phi_{X^\bullet} =$$
$$[h_{\bullet X} \otimes (^*X)^*][^*X \otimes \sigma_{(X^*)^\bullet,(^\bullet X)^\bullet}][\theta_{\bullet X} \otimes \theta_{(^\bullet X)^\bullet} \otimes {}^*(X^*)]$$
$$[\sigma^2 \otimes {}^*(X^*)][\epsilon_{\bullet X} \otimes {}^*(X^*)].$$

It thus follows that if θ satisfies the balancing axiom

$$\theta_{A \otimes B} = [\theta_A \otimes \theta_B]\sigma_{A,B}\sigma_{B,A},$$

then ϕ defined in the theorem gives a sovereign structure on the category for the given right duals and left duals obtained by "twisting" with the braiding.

The key to the reverse implication is to consider in detail the condition that ϕ be a *monoidal* natural transformation. Let

$$b_{X,Y} : (X \otimes Y)^* \to Y^* \otimes X^*$$

be the canonical isomorphism which makes $(-)^*$ into a monoidal functor. In this case, the condition that b be the structure maps for the monoidal functor is equivalent to the condition

$$\eta_{X\otimes Y}[b \otimes X \otimes Y] = \eta_Y[Y^* \otimes \eta_X \otimes X^*]$$

and a similar condition relating ϵ and b^{-1}.

Composing both sides of this equation with σ and applying the naturality of σ to both sides shows that

$$h_{X\otimes Y}[X \otimes Y \otimes B] = h_X[X \otimes h_Y \otimes {}^*X][X \otimes Y \otimes \sigma^2],$$

and a similar calculation for the condition on ϵ and e shows that the structure map for ${}^*(-)$ as a monoidal functor is $b\sigma^{-2}$. The condition that ϕ be a monoidal natural transformation becomes

$$b[\phi \otimes \phi] = \phi b \sigma^{-2}$$

or equivalently,

$$b_{X,Y}[\phi_X \otimes \phi_Y]\sigma^2 b_{X,Y}^{-1} = \phi_{X\otimes Y}.$$

Now recalling the definition of θ in terms of ϕ, and calculating $\theta_{X\otimes Y}$ by substituting the left-hand side of the last equation for $\phi_{X\otimes Y}$, applying the defining property of b and using the naturality and invertibility of the braiding, we obtain the balancing condition for θ as defined in terms of ϕ.

Thus we establish the lemma and the theorem. $\square$ $\square$

In the case of categories of modules over a bialgebra A, the structures discussed in this chapter correspond to the additional structures discussed at the end of the previous chapter. We state without proof:

Theorem 5.16 *If A is a bialgebra over K, then the following implications hold for the category A-**mod** with the induced monoidal structure of Example 3.7:*

1. *If A is a Hopf algebra , then A-**mod** has right (and left) duals.*

2. *If A is triangular , then A-**mod** is symmetric.*

3. *If A is quasi-triangular, then A-**mod** is braided.*

4. *If A is ribbon, then A-**mod** is ribbon (tortile).*

In the first and last cases, the module structure on on the dual space is induced by $[a \cdot f](b) = f(S(a) \cdot b)$. In the last three, the braiding is given by $x \otimes y \mapsto R \cdot [y \otimes x]$. In the last, the balancing map θ is multiplication by v.

If one considers the dual setting, and deals with categories of comodules, a similar result holds, once the algebraic notions are also dualized. In this case one also has a converse by way of "Tannaka-Krein" reconstruction (cf. [47, 54]).

Chapter 6

Knot Polynomials

The original discovery of the interplay between monoidal category theory and classical knot theory was motivated by a desire to better understand Laurent-polynomial valued invariants of knots and links discovered by Jones [29], HOMFLY [42] (cf. also Przytycki and Traczek [43]), Brandt, Lickorish and Millet [10], and Kauffman [32].

Although the actual constructions given by the authors cited vary greatly in detail, the description of the invariants can be reduced to combinatorial relations known as *skein relations*.[1] To describe a link invariant by skein relations, we let $\mathcal{L}$ (resp. $\mathcal{L}_f$) be the free R-module with the ambient isotopy classes of classical links (resp. framed links) as its basis. A skein relation is then specified by giving a formal linear combination of tangles (resp. framed tangles), usually with two inputs and two outputs, but in any event with the same intersection with the boundary of the cube. Any skein relation specifies a submodule of $\mathcal{L}$ (or $\mathcal{L}_f$) by taking as generators *all* linear combinations of ambient isotopy classes of links whose summands admit representatives which contain

---

[1]Some researchers, notably Przytycki and Hoste have taken the notion of skein relations very seriously, and have developped a species of algebraic topology based on considering the space linear combinations of ambient isotopy classes of links in a 3-manifold modulo (a) given skein relation(s) as an invariant of the space. Alas, discussion of these interesting developments is beyond the scope of this present work.

the tangles of the summands inside some given cube but are identical outside the given cube, with the coefficients of the corresponding tangles from the skein relation.

A skein relation (or family of skein relations) then defines an R-valued invariant of links (or framed links) if the quotient module is free of rank 1. (One can always normalize so that the equivalence class of the unknot is identified with $1 \in R$.)

The existence theorems for the Jones and HOMFLY polynomials may then be stated as

Theorem 6.1 (Jones) *Let* $R = \mathbb{Z}[i, t^{\frac{1}{2}}, t^{-\frac{1}{2}}]$, *then the skein relation*

$$it^{-1}T_+ - itT_- + (t^{\frac{1}{2}} - t^{-\frac{1}{2}})T_0$$

determines a unique R-valued ambient isotopy invariant of classical links.

and

Theorem 6.2 (HOMFLY,P-T) *Let* $R = \mathbb{Z}[x, x^{-1}, z, z^{-1}]$, *then the skein relation*

$$xT_+ - x^{-1}T_- + zT_0$$

determines a unique R-valued ambient isotopy invariant of classical links.

In each of the theorem the tangles T_+, T_- and T_0 are two strand tangles with both strands oriented downward, with T_+ being the positive crossing, T_- the negative crossing, and T_0 having no crossings, when each is projected onto the back wall of the cube containing the tangle. (See Figure 2.2 for our convention concerning crossing signs.)

The original discovery of Jones [29] and the constructions of Ocneanu and of Freyd and Yetter for the HOMFLY polynomial [42] all rely upon the following theorems to reduce the construction of an ambient isotopy invariant of classical links to the construction of a suitable family of representations of Artin's braid groups:

Theorem 6.3 (Alexander [1]) *Every ambient isotopy class of oriented links can be represented by a closed braid, that is, by a link obtained from a geometric braid with all strands oriented downward by joining corresponding strands at the top and bottom of the geometric braid by a family of non-intersecting arcs all lying in a plane.*

Theorem 6.4 (Markov [41], Birman [9]) *Two braids, $\beta_1 \in B_m$ and $\beta_2 \in B_n$, have isotopic closures if one can be obtained from the other by a sequence of moves of the following types:*

$$\beta \longleftrightarrow \gamma\beta\gamma^{-1} \text{ for some } \gamma \text{ in the same braid group as } \beta$$

$$\beta \longleftrightarrow [\beta \otimes 1]\sigma_n^{\pm 1}$$

where $\beta \in B_n$ and the notations are as in the previous chapter.

In Jones's original work [29] and Ocneanu's construction of the HOMFLY polynomial [42], the skein relation passes easily to a relation on a family of linear representations of the braid groups, and a family of traces on the representations satisfying suitable properties abstracted from the second move in Markov's Theorem gives rise to the link invariant.

In the construction of Freyd and Yetter of the HOMFLY polynomial [42], the skein relation, together with the second Markov move, generate a submodule of the direct sum of the group algebras of the braid groups, and it is shown that the quotient module is cyclic of rank one, and that any two conjugate braids have the same image in the quotient.

Neither of these approaches is successful in accounting for the invariants of Brandt, Lickorish and Millet [10], or Kauffman [32], because the skein relations for these invariants (which use unoriented diagrams) involve a fourth tangle, T_∞, whose projection is crossing free, but has the two top inputs (and two bottom inputs) connected to each other by an arc. The monoidal categories of tangles discussed in the next chapter were first defined by the author [59] in response to this difficulty.[2]

---

[2] Categories of tangles were discovered independently and slightly later by Turaev [53].

Chapter 7

Categories of Tangles

We now return to our topological motivation for introducing the particular concepts from category theory. Not only do tangles, oriented tangles and framed tangles form categories with composition given by "paste-and-rescale"; in fact, they form ribbon categories.

Just as we had defined the composition of two arrows in **Tang**, **OTang** or **FrTang** by attaching two copies of the cube $\mathbb{I}^3$ bottom-face to top-face and rescaling, we can use the same pasting and rescaling procedure along the second coördinate to define a monoidal structure on any of these categories.

To discuss the structure maps, it is helpful to generalize slightly the notion of geometric braid introduced in connection with the Artin braid groups:

Definition 7.1 *A geometric tangle is a geometric braid if it is the image of a map of the form* $T : \{1, \ldots, n\} \times \mathbb{I} \to \mathbb{I}^3$ *satisfying* $p_3(T(i,t)) = t$.

Lemma 7.2 *In any of* **Tang**, **OTang** *or* **FrTang**, *any arrow that admits a representative whose underlying geometric tangle is a geometric braid is an isomorphism.*

proof: Given an arrow with a representative geometric braid, the inverse arrow is given by the mirror image in a horizontal plane with the

87

orientations (if any) reversed. It is easy to verify (using ambient iso-topies corresponding to Reidemeister moves of type $\Omega.2$ in the projection onto the back wall) that these arrows are inverses in **Tang** and **OTang**. For **FrTang** one must also note that the mirror-imaged framing twists undo those in the given tangle. □

In fact, all of the structural natural isomorphisms in the categories of tangles are of this form.

The monoidal structure on any of our categories of tangles is then given by

Definition 7.3 Monoidal structures on categories of tangles

The monoidal product $T_1 \otimes T_2$ of two tangles T_1, T_2 is given by the map induced on the disjoint union of the underlying 1-manifolds, by the composition of $T_1 \coprod T_2$ with the map $\gamma_3 : \mathbb{I}^3 \coprod \mathbb{I}^3 \to \mathbb{I}^3$ given by

$$(x, y, z) \mapsto (\frac{x}{2}, y, z) \text{ for elements of the first summand, and}$$

$$(x, y, z) \mapsto (\frac{x+1}{2}, y, z) \text{ for elements of the second summand}$$

with orientations and framings (if any) given in the obvious manner.
I is the empty subset of $\mathbb{I}$.

The structural natural transformations have components given by those geometric braids with all component maps constant in the y-coördinate, which are shown schematically in Figure 7.1.

The two directions around the pentagon and triangle coherence con-ditions are shown schematically in Figure 7.2. It is easy to see that there are ambient isotopies which implement the equality between the two directions around each coherence diagram.

In fact, we can say more.

Proposition 7.4 *The categories* **Tang**, **OTang** *and* **FrTang** *are all ribbon categories when equipped with the braiding, twist map and du-alities given as follows: The braiding $\sigma_{A,B}$ is given by composing the*

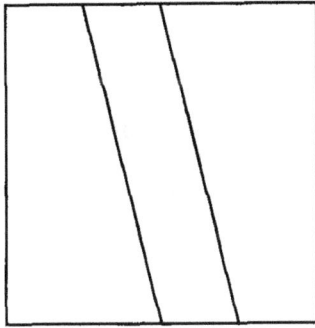

α

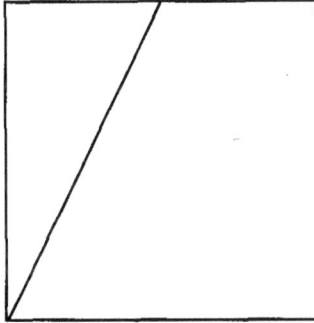

λ

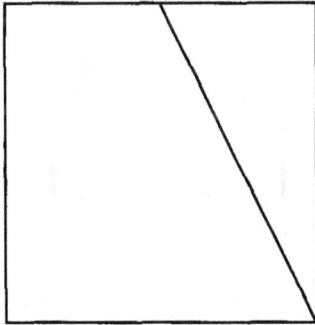

ρ

Figure 7.1: Structural Natural Transformations for Categories of Tangles

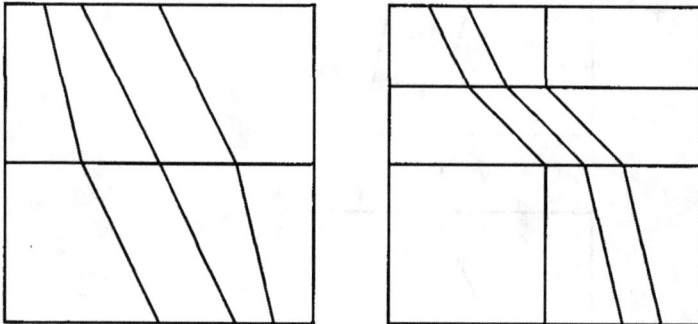

pentagon

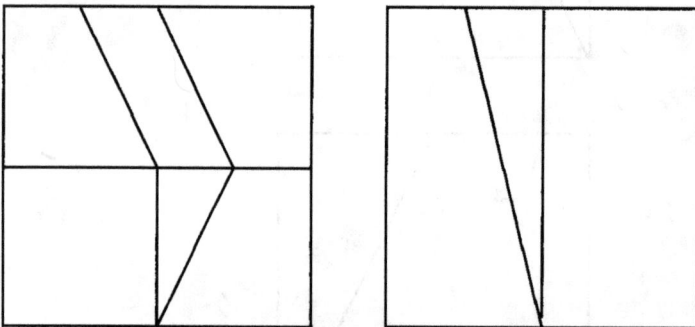

triangle

Figure 7.2: Coherence Conditions for Categories of Tangles

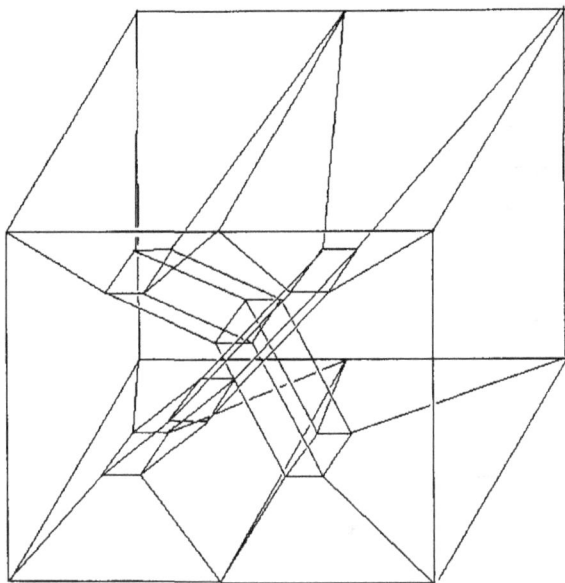

Figure 7.3: Inclusion Inducing the Braiding

map $Id_A \coprod Id_B : A \times \mathbb{I} \coprod B \times \mathbb{I} \to \mathbb{I}^3 \coprod \mathbb{I}^3$ with the level-preserving PL inclusion of $\mathbb{I}^3 \coprod \mathbb{I}^3$ into $\mathbb{I}^3$ shown in Figure 7.3. The twist map θ_A is given by composing the map naming Id_A with a level-preserving PL homeomorphism from $\mathbb{I}^3$ to itself that rotates the bottom level through $360°$ clockwise (full clockwise twist). Right dual objects are given by reflection in the x-coördinate, with orientations reversed in the case of **OTang** *and* **FrTang**, *with structure maps η_A and ϵ_A given by composing the map naming Id_A with the PL inclusion of $\mathbb{I}^3$ into $\mathbb{I}^3$ shown in Figure 7.4.*

proof: That the maps given as units and counits for right duals satisfy the required equations (and thus that the given objects *are* right duals) follows immediately from isotopies which are instances of Reidemeister's move $\Delta.\pi.1$ when viewed in projection onto the back wall ($\mathbb{I} \times 1 \times \mathbb{I} \subset \mathbb{I}^3$).

The naturality condition for σ can be verified using isotopies that are generated by composing the isotopy implementing the condition

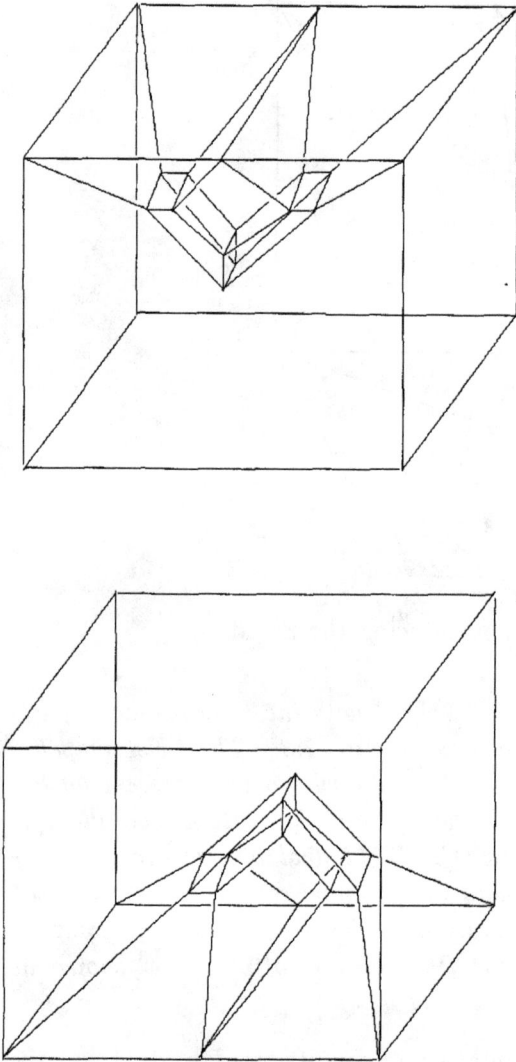

Figure 7.4: Inclusions Inducing the Unit and Counit for Right-Dual
Objects

defining identity maps in categories of tangles with one of the inclusion of Figure 7.5.

The hexagonal coherence condition for the braiding is induced by an isotopy which "straightens" the crossings (viewed front to back).

Once it is observed that all full twists are ambient isotopic rel boundary, it is easy to see that the defining condition of a balancing holds: a full twist on each tensorand followed by two instances of the braiding accomplishes a full twist on the tensor product, as shown schematically in Figure 7.6.

Similarly, it is easy to see that a full clockwise twist applied to an object's dual is the dual of a full clockwise twist applied to the object, verifying the condition relating the balancing and duality. □

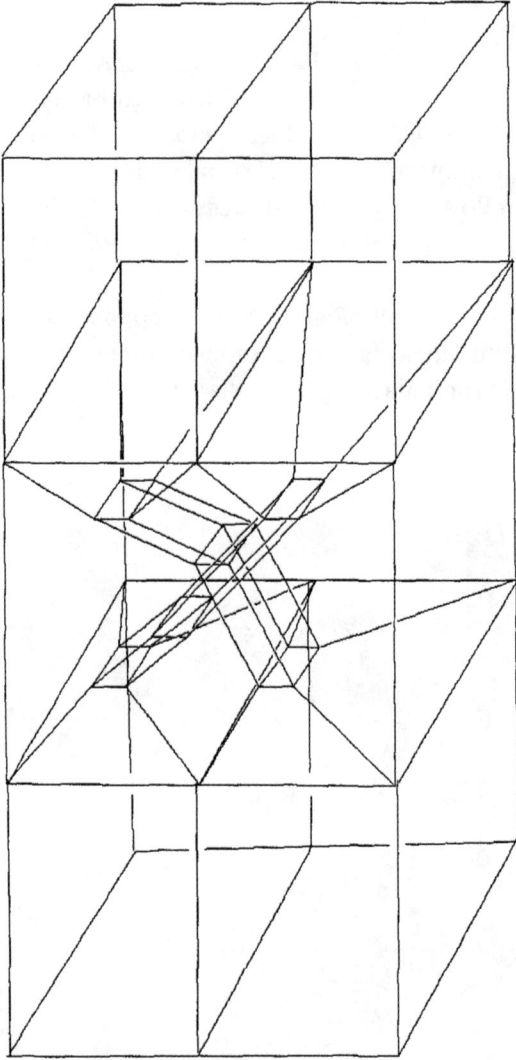

Figure 7.5: Inclusion Inducing Naturality Isotopies for Braiding

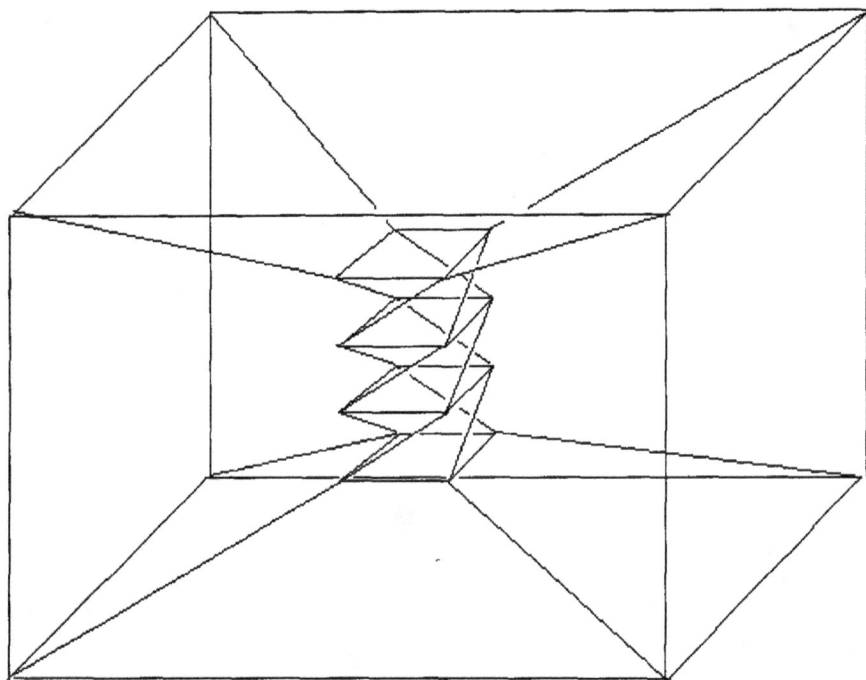

Figure 7.6: Schematic of Balancing for Categories of Tangles

Chapter 8

Smooth Tangles and PL Tangles

Thus far we have handled everything in the PL setting, even going so far as to translate the more naturally smooth notion of framings for knots, links and tangles into that setting. Now that we have sufficient categorical machinery set up, we wish to take time to see that the two approaches give rise to equivalent formulations of tangle theory. By smooth, we mean C^∞, although similar proofs to those given below would give corresponding results for C^k with $k \geq 1$.

Our method will be to construct a number of different categories of tangles and framed tangles, and show that they are all ribbon equivalent to one another (that is, monoidally equivalent by a monoidal functor which preserves the braiding and balancing – duals come along for free). **Tang**, **OTang** and **FrTang** will, of course denote the PL versions constructed in the previous chapter. As in the PL case, we assume when dealing with framed tangles that both the underlying manifold of the tangle and the target $\mathbb{I}^3$ are oriented.

The key to this chapter is to introduce categories of piecewise smooth tangles, oriented tangles and framed tangles, which can be readily related to the categories of the last chapter and to categories constructed by smooth methods. In fact, the categories **Tang**$_{PS}$, **OTang**$_{PS}$ and

FrTang$_{PS}$ are defined in exactly the same way as those of the previous chapter, but with piecewise linear inclusions and isotopies replaced by piecewise smooth ones. To make this more precise, we make:

Definition 8.1 *A piecewise smooth map is a continuous map for which there exists a triangulation of the source such that the restriction of the map to each simplex is C^∞.*

Thus, any PL map is piecewise smooth. Conversely,

Theorem 8.2 *If $F : X \to Y$ is a piecewise smooth embedding where $Y = \mathbb{R}^3$ (resp. a piecewise smooth tangle with PL source and target), and X is a surface, then F can be approximated by a PL embedding $\hat{F}$ which is, moreover, piecewise smooth ambient isotopic (resp. ambient isotopic rel boundary) to F.*

As one technique of proof for these approximation theorems involves lemmas we will need later to prove some results about categories of framed tangles, we will not rely on any "big theorems" of differential topology. Instead, we will give a rather hands-on account using only elementary facts from differential topology found in standard texts (e.g. Guilleman and Pollack [27] or Spivak [50]). One not-so-big theorem, which would be very convenient — the result of Alexander which states that any homeomorphism or diffeomorphism which fixes an open region is ambient isotopic to the identity — will not be available to us, since we need relative versions in a cube. The inversions which give the proof of Alexander's result do not respect the cube.

Definition 8.3 *A* smooth ramp function from (a, b) to (c, d) *is a smooth function* $\phi : \mathbb{R} \to \mathbb{R}$ *satisfying*

- *ϕ is weakly monotone*
- *$\phi(a) = b$*
- *$\phi(c) = d$*

- $\phi'(x) = 0$ *for all* $x \notin (a, c)$.

A smooth bump function for $[a, b] \subset (c, d)$ with inner value y and outer value z *is a smooth function* β *such that*

- $\beta(x) = y$ *for all* $x \in [a, b]$

- $\beta(x) = z$ *for all* $x \notin (c, d)$

- β *is monotone on* $[c, a]$

- β *is monotone on* $[b, c]$.

For our purposes, it is desirable to observe that there are families of ramp functions and bump functions which vary smoothly in the parameters a, b, c, d or a, b, c, d, y, z. More precisely,

Proposition 8.4 *There exists a smooth function*

$$\Phi(a, b, c, d, x) \text{ (resp. } B(a, b, c, d, y, z, x))$$

such that for fixed a, b, c, d (resp. a, b, c, d, y, z) the function $\phi(x) = \Phi(a, b, c, d, x)$ for $a < c$ (resp. $\beta(x) = B(a, b, c, d, y, z, x)$ for $c < a < b < d$) is a smooth ramp function from (a, b) to (c, d) (resp. a smooth bump function for $[a, b] \subset (c, d)$ with inner value y and outer value z).

proof: We begin by defining $\Phi(0, 0, 1, 1, x)$ to be the standard smooth ramp function $\varphi(x)$, found for example in Spivak [50]: let

$$f(x) = \begin{cases} e^{-\frac{1}{x^2}} & \text{if } x > 0 \\ 0 & \text{if } x \leq 0. \end{cases}$$

Then

$$\varphi(x) = \frac{\int_0^x f(\xi) f(1 - \xi) d\xi}{\int_0^1 f(\xi) f(1 - \xi) d\xi}.$$

Now, let $\Phi(a,b,c,d,x) = (d-b)\varphi(\frac{x-a}{c-a}) + b$. Plainly Φ has all the desired properties.

Now, define B by

$$B(a,b,c,d,y,z,x) = \begin{cases} \Phi(a,z,c,y,x) & \text{if } x \leq \frac{a+b}{2} \\ \Phi(b,y,d,z,x) & \text{if } x > \frac{a+b}{2}. \end{cases}$$

Again, it is immediate that the function has the desired properties. $\square$

Moreover,

Theorem 8.5 *If H is a piecewise smooth ambient isotopy rel boundary between PL tangles F and G, then H can be approximated by a PL ambient isotopy rel boundary.*

proof: Now, a piecewise smooth (resp. PL) ambient isotopy rel boundary is given by a piecewise smooth (resp. PL) map $\phi : \mathbb{I} \times \mathbb{I}^3 \to \mathbb{I}^3$ such that for all t $\phi_t(-) = \phi(t,-)$ has a piecewise smooth (resp. PL) inverse, and $\phi_t|_{\partial \mathbb{I}^3} = Id_{\partial \mathbb{I}^3}$.

It is easy to see that ϕ is a piecewise smooth (resp. PL) ambient isotopy if and only if $(p_1, \phi) : \mathbb{I} \times \mathbb{I}^3 \to \mathbb{I} \times \mathbb{I}^3$ is a piecewise smooth (resp. PL) homeomorphism.

First, observe that the ε-neighborhood lemma of [27] applies to embedded compact closed submanifolds with boundary as well as to embedded compact submanifolds.[1] Thus, we can apply the ε-neighborhood theorem to each of the (finitely many) simplexes τ of a triangulation with respect to which (p_1, ϕ) is piecewise smooth, to obtain a neighborhood of each simplex of radius ε_τ on which the image of the normal bundle retracts by the nearest-neighbor map onto the simplex.

Let ε be the minimum of the ε_τ's. For any simplex τ let τ_ε denote the ε neighborhood of the image of τ under (p_1, ϕ), and let N_τ be the image of the regular neighborhood of τ under (p_1, ϕ). Now, choose a

---

[1]The stronger version in which the bound is given on a non-compact submanifold by a continuous positive function can be applied to the smooth embedding of an open neighborhood of the manifold with boundary, then ε chosen to be the minimum of this function on the compact closed submanifold.

subdivision of the triangulation such that every "secant simplex" s, that is, every linear simplex s whose vertices are the images under (p_1, ϕ) of vertices of a simplex σ of the subdivision, lies in $\tau_\varepsilon \cap N_\tau$, where τ is the simplex of the original triangulation in which σ lies. Now, let $\Psi : \mathbb{I} \times \mathbb{I}^3 \to \mathbb{I} \times \mathbb{I}^3$ be the map obtained by linearly mapping each simplex of the subdivision to the corresponding secant simplex. First, observe that since (p_1, ϕ) is linear in its first coördinate, the first coördinate of Ψ is also p_1, so that we have $\Psi = (p_1, \psi)$. Likewise, (p_1, ϕ) restricted to $\mathbb{I} \times \partial \mathbb{I}^3$ is the identity, and thus linear. Thus Ψ agrees with (p_1, ϕ) on the boundary.

We need to show that Ψ is invertible. It is immediate by construction that Ψ is surjective. To see that it is one-to-one, we proceed considering the skeleta of the subdivision. It is immediate that Ψ is one-to-one on the 0-skeleton (vertices) since (p_1, ϕ) is one-to-one. On higher skeleta, it is immediate that the restriction of Ψ to any simplex of the subdivision is one-to-one, so we only need see that the images of simplexes intersect only when one is the face of another. Suppose we had an intersection not of this form. The two points with the same image cannot lie in the image of the same simplex of the original triangulation, since the secant simplexes all lie in the ε-neighborhood, and thus retract onto the image of the simplex under (p_1, ϕ). Likewise, they cannot lie in images of different simplexes, since then they would have to lie in the image of the regular neighborhood of each, and thus lie in a common face, leading to the previous contradiction.

Thus ψ such that $\Psi = (p_1, \psi)$ is the desired PL ambient isotopy rel boundary. $\square$

It therefore follows that the obvious inclusion of sets of PL tangles into sets of piecewise smooth tangles descends to a map on equivalence classes, which plainly induces functors

$$J : \mathbf{Tang} \to \mathbf{Tang}_{PS}$$

$$J : \mathbf{OTang} \to \mathbf{OTang}_{PS}$$

$$J : \mathbf{FrTang} \to \mathbf{FrTang}_{PS}.$$

In fact, the same approximation theorems show that these functors are actually isomorphisms of categories. The fact that the structure maps for the piecewise smooth categories are represented by PL tangles then shows that these are ribbon equivalences as well.

The first difficulty in the program of showing smooth categories of tangles to be equivalent to their corresponding PL version comes from the fact that although PL and piecewise smooth embeddings will "glue" end-to-end with no more data than the points (signed or framed as needed) at which they are to be glued, one must do more work to glue smooth embeddings end-to-end.

This is resolved by another approximation theorem by which any piecewise smooth map, smooth at its boundary, can be approximated by smooth maps. We can glue to get a piecewise smooth embedding, which is smooth except near each gluing point, then locally approximate by new smooth embeddings near these points to define the composition. Similarly, smooth embeddings which are smooth ambient isotopic (rel boundary) are *a fortiori* piecewise smooth ambient isotopic (rel boundary), but by the approximation theorem the converse holds.

Thus, in the case of tangles and oriented tangles, we are in the same position as we were for relating the PL and piecewise smooth categories: there are functors

$$L : \mathbf{Tang}_{C^\infty} \to \mathbf{Tang}_{PS}$$

$$L : \mathbf{OTang}_{C^\infty} \to \mathbf{OTang}_{PS}$$

which are, in fact, isomorphisms of ribbon categories.

Even for "framed tangles" the same situation applies, but for reasons which will become clear, we will now revert to Shum's name of "ribbon tangles" [48] for the smooth version of our PL and piecewise smooth tangles, and denote the category by **RibTang**$_{C\infty}$. Thus by the same argument, there is an isomorphism of ribbon categories

$$L : \mathbf{RibTang}_{C\infty} \to \mathbf{FrTang}_{PS}.$$

The reason for our name change is that the arrows in this category are smooth embeddings of "ribbons", that is, of manifolds with boundary of the form $X \times \mathbb{I}$, where X is a disjoint union of components diffeomorphic to S^1 and $\mathbb{I}$. In the smooth setting, we can use framed tangle to describe precisely what it sounds like: a tangle equipped with a framing of the normal bundle.

We denote the category of smooth framed tangles in this sense by **FrTang**$_{C\infty}$. The one equivalence (and it is only an equivalence, not an isomorphism) which does not follow from an approximation theorem is

Theorem 8.6 FrTang$_{C\infty}$ *is ribbon equivalent to* **RibTang**$_{C\infty}$.

proof: First, recall that the underlying manifolds and ambient $\mathbb{I}^3$ are oriented, and thus the specification of a framing of the normal bundle of the tangle can be reduced to the specification of a normal vectorfield on the tangle. In specifying this normal vectorfield, we are really specifying a map from a one-jet neighborhood (in $\mathbb{R}^2$) of the underlying manifold of the tangle.

To describe this more precisely, we can follow Goryunov [26] in regarding framed knots, links and tangles not as ordinary knots, links or tangles with additional structure, but as equivalence classes of mappings from open annular neighborhoods U of S^1 or disjoint unions of such (and of the same with open rectangular neighborhoods U of $\mathbb{I} = \mathbb{I} \times \{0\} \subset \mathbb{I} \times \mathbb{R}$ in the case of tangles) into $\mathbb{R}^3$ (or $\mathbb{I}^3$ in the case of tangles). We interrupt the proof briefly to recall the relevant definitions:

Definition 8.7 *Given an oriented space X of the form S^1, $\mathbb{I}$, $\{*\}$, or a disjoint union of such spaces (all of the same dimension), a ribbon neighborhood of the space is an oriented space of the form $U = X \times (-a, a)$ $(a > 0)$. We identify X with $X \otimes \{0\} \subset U$.* [2]

---

[2] We prefer the suggestive "ribbon neighborhoods" to the more classical "open bicollar neighborhoods".

Observe that any mapping $g : U \to M$ induces a mapping Tg from $i^*(T\mathbb{R}^{i+1})$ to $T(M)$. (Here i is the inclusion of X into U, and although the observation holds for more general target manifolds with boundary, in our case M is $\mathbb{R}^3$, $\mathbb{I}^3$ or $\mathbb{I}^2$.)

For ease in the link and tangle settings, we may specify in advance a countable (ordered) family of disjoint circles and intervals

$$\{S_1, \ldots S_k, \ldots, \mathbb{I}_1, \ldots \mathbb{I}_l, \ldots\}$$

and consider mappings from disjoint neighborhoods of finite unions of families of components, which are initial segments of the components of each type.

Definition 8.8 *Two mappings $g_i : U_i \to \mathbb{R}^3$ (resp. $g_i : U_i \to \mathbb{I}^3$) $i = 1, 2$ are* equivalent *if the U_i's are ribbon neighborhoods of the same disjoint union of circles (resp. disjoint union of circles and intervals) X and the mappings $Tg_i : i^*(T\mathbb{R}^2) \to T\mathbb{R}^3$ coincide on X.*

Observe that an equivalence class of embeddings of a ribbon neighborhood specifies a framed link, since we can take the vectorfield to be the image of the unit normal vectorfield to the underlying manifold under Tg. Conversely, any framing (given by its "first" vectorfield) specifies an equivalence class of mappings from ribbon neighborhoods to $\mathbb{I}^3$ — namely, that equivalence class for which the normal vectorfield is the image under Tg of the unit normal vectorfield to the underlying manifold.

Actually, a little work is needed here: we need to know that there is an embedding of a ribbon neighborhood for which any given normal vectorfield is the image of the unit normal vectorfield. This, however, is easy: apply the exponential map associated to the Euclidean metric on $\mathbb{I}^3$ to obtain a map from a ribbon neighborhood. Now, linearly scale the parameterization normal to the tangle to obtain a map g for which the image of the unit normal vectors are the given tangent vectors. It only remains to restrict g to a smaller ribbon neighborhood on which it is an embedding.

By the immersion theorem (cf., for example Guilleman and Pollack [27]), the map is an immersion on some neighborhood of the tangle, and by compactness this may be chosen to be a ribbon neighborhood. Similarly, we can restrict to a ribbon neighborhood on which the map it one-to-one, since otherwise we can construct sequences of points $\{x_n\}$ and $\{y_n\}$ such that

- $x_n \neq y_n$

- $g(x_n) = g(y_n)$

- $d(x_n, X) < \frac{1}{n}$

- $d(y_n, X) < \frac{1}{n}$

where X is the underlying manifold of the tangle, and distance is measured in the ribbon neighborhood. Restricting to the final subsequences which lie in some $X \times [-\frac{1}{n}, \frac{1}{n}]$, by compactness we may assume w.l.o.g. that both sequences are convergent. Now, it is plain by construction that the limits x and y lie in the tangle $(g(X))$, and that $g(x) = g(y)$. However this is a contradiction, since either $x \neq y$, violating the embedding condition in the definition of tangles, or $x = y$, in which case the sequences converging to them violate the fact that immersions are local embeddings.

proof of Theorem 8.6 continued:
This being said, it is clear how to obtain a functor

$$T : \mathbf{RibTang}_{C^\infty} \to \mathbf{FrTang}_{C^\infty}.$$

Whether on objects or on arrows, simply take the equivalence class represented by the map. (One needs to observe, since ribbon tangles were defined via maps from $X \times \mathbb{I}$, that the definition of smoothness for maps on manifolds with boundary allows us to extend this to a map on an open neighborhood of this, which by compactness of X may be assumed to be of the form $X \times (-a, 1+a)$, and that all such extensions, when restricted to $X \times (-a, a)$, are equivalent in the sense above.)

We thus need to construct a functor in the reverse direction and the appropriate natural isomorphisms to establish the theorem. Most of the work has already been done in the preceding discussion. The only real problem is how to choose a particular neighborhood on which to define the map, and then rescale so that mapping of normal vectors on the tangle is unchanged, but the embedding is defined on $X \times \mathbb{I}$. The only crucial thing about this choice is the choice on objects, since we have

Lemma 8.9 *If $g_1, g_2 : X \times \mathbb{I} \to \mathbb{I}^3$ are ribbon tangles satisfying*

- $g_1|_{X \times \{0\}} = g_2|_{X \times \{0\}}$

- $g_1|_{\partial X \times \mathbb{I}} = g_2|_{\partial X \times \mathbb{I}}$

- $T g_1|_{X \times \{0\}} = T g_2|_{X \times \{0\}}$

then g_1 and g_2 are ambient isotopic rel boundary.

   The same statement holds when the ribbon tangles are replaced with maps $g_1, g_2 : S \times \mathbb{I} \to \mathbb{I}^2$ and the second condition is dropped.

We will defer the proof of Lemma 8.9 until we are done proving Theorem 8.6.

   On objects, given a (normal) vectorfield v_s on a finite set of points S in $(0,1)^2 \subset \mathbb{I}^2$, we need to specify an embedding of $S \times \mathbb{I}$ for which the image of the unit tangent vector at $(s, 0)$ is v_s. Now, for each point in S, let $d_s = \frac{1}{4} \min(d(s, \partial \mathbb{I}^2), d(s, S \setminus \{s\}))$, and define the embedding g_v by

$$(s, t) \mapsto s + \frac{v_s}{k_s} \phi_s(t)$$

where ϕ_s is a smooth monotone increasing function with the properties

- $\phi_s(0) = 0$

- $\phi_s(1) = \frac{d_s}{\|v_s\|}$

- $\phi'_s(0) = 1$.

We will actually construct functions satisfying this kind of requirements in the proof of Lemma 8.12 where additional requirements will
be needed, so we defer the construction until then.

The functor $R : \mathbf{FrTang}_{C^\infty} \to \mathbf{RibTang}_{C^\infty}$ is then given on objects
by mapping signed sets of points to the embeddings just constructed.

On arrows, we will use the same construction, but with $\phi_s(t)$ replaced with a smooth non-negative function $\phi(x, t)$ satisfying

- $\phi(x, 0) = 0$

- $\frac{\partial \phi}{\partial t}(x, 0) = 1$

- If $s \in \partial X$, then $\phi(s, t) = \frac{d_s}{\|v_s\|}$ as above

- $(x, t) \mapsto x + \frac{v_x}{\phi}(x, t)$ is an embedding.

To avoid some difficulties later, first perform an ambient isotopy
rel boundary so that the tangle intersects a δ-neighborhood of $\partial \mathbb{I}^3$ in a
family of vertical line segments (any parameterization will do).

To see that such $\phi(x, t)$ exist first apply the ϵ-neighborhood theorem
[27] to the tangle. Now, let

$$X_\epsilon = X \setminus N_\epsilon(\partial X)$$

where N_ϵ is the ϵ-neighborhood in $\mathbb{I}^3$, and replace ϵ with

$$\varepsilon = \frac{d(X_\epsilon, \partial \mathbb{I}^3)}{3} \leq \frac{\epsilon}{3}.$$

Let $\vartheta = \min(\delta, \varepsilon)$. We can now form a neighborhood of X as follows:
let X^ε be the closed ε neighborhood of the tangle X, and let N_s be the
closed neighborhood of $s = (x, y, z)$ of the form $\overline{P}_{d_s}(x, y) \times (\overline{B}_\vartheta(z) \cup \mathbb{I})$.
Let $\delta(x)$ be defined by

$$\delta(x) = \sup\{\, m \mid t \leq m \text{ implies } x + tv_x \in X^\varepsilon \cup \bigcup_s N_s \,\}$$

and let $d(x)$ be any smooth function chosen so that $d(s) = \delta(s)$ for $s \in \partial X$, and $0 < d(x) \leq \delta(x)$ for all $x \in X$.

Now, by construction, the exponential map $(x, v) \mapsto x + v$ restricted to $\{\, (x, v) \mid x \in X, \ v \text{ is normal to X}, \ \|v\| \leq d(x) \,\}$ is an embedding. The ribbon map constructed is simply the composite of this embedding with the embedding $(x, t) \mapsto (x, \frac{v_x}{\phi}(x, t))$ from $X \times \mathbb{I}$ to $N(X)$, where $\phi(x, t)$ is a smooth function satisfying

- $\phi(x, 0) = 0$

- $\frac{\partial \phi}{\partial t}(x, 0) = 1$

- If $s \in \partial X$, then $\phi(s, t) = \frac{d_s}{\|v_s\|}$, as above

- $\phi(x, 1) = \frac{d(x)}{\|v_x\|}$.

Again, the construction in Lemma 8.12 will provide the necessary function.

Now, by Lemma 8.9 any different choices of δ, ϵ and $\delta(x)$ will give ambient isotopic ribbons. It is trivial to see that composition, identities, $\otimes$, I and all of the structure maps are preserved. Thus we have the desired ribbon functor R.

It is immediate from the construction that $TR = Id_{\mathbf{FrTang}_{C^\infty}}$.

To construct the required natural isomorphism

$$\psi : RT \Rightarrow Id_{\mathbf{RibTang}_{C^\infty}},$$

let H be the ambient isotopy provided by the isotopy of the second statement of Lemma 8.9 from an object $h : S \times \mathbb{I} \to \mathbb{I}^2$ to the map naming the object $T(R(h))$ constructed above. The component ψ_h of the natural isomorphism is then named by

$$\psi_h(x, t, \tau) = (H(h(x, t), \tau), \tau)$$

and its inverse named by

$$(x, t, \tau) \mapsto (H(h(x, t), 1 - \tau), \tau).$$

Both the inverse and naturality conditions follow immediately from the application of the lemma. Thus, once we establish the lemma, the theorem is proven. □

proof of Lemma 8.9 We proceed by constructing a sequence of four ambient isotopies, of which the first two adjust the embeddings along the source and target so that the map on tangent spaces coincides there as well, and the third of which shrinks the ribbon tangle close enough to the edges where the tangent maps coincide, so that the fourth can complete the desired isotopy using

Lemma 8.10 *There exists an $\epsilon > 0$ such that if*

$$\psi(x, y) = (x + a(x, y), y + b(x, y), c(x, y) + \frac{1}{2})$$

is an embedding of $\mathbb{I}^2$ in $\mathbb{I}^3$ and

$$|a|, |b|, |c|, |a_x|, |b_x|, |c_x|, |a_y|, |b_y|, |c_y| < \epsilon$$

then ψ is ambient isotopic to the map $(x, y) \mapsto (x, y, \frac{1}{2})$. Moreover, the isotopy may be chosen so that all points where $a = b = c = 0$ are fixed.

proof: First, observe that there exist $\delta, \eta > 0$ such that ψ extends to an embedding of $[-\delta, 1 + \delta]^2$ into $[-\eta, 1 + \eta]^3$. Now, choose a smooth function $\phi : [-\eta, 1 + \eta]^3 \times \mathbb{I} \to \mathbb{I}$ such that

- $\phi|_{N \times \mathbb{I}} = 0$ for N a neighborhood of $\partial[-\eta, 1 + \eta]^3$,

- $\phi|_{M \times [c,1]} = 1$ for some $0 < c < 1$ and M a neighborhood of $\mathbb{I}^2 \times \{\frac{1}{2}\}$,

- $\phi|_{[-\eta,1+\eta]^3 \times [0,k]} = p_{[-\eta,1+\eta]^3}$ for some $0 < k < c$.

(We can construct the desired ϕ as a product of smooth bump functions in each of the first three variables x, y, z and a smooth ramp function in the fourth variable t.)

Now, let

$$\mu = \max(\ \max_{\substack{a \in \{x,y,z\} \\ (x,y,z,t) \in [-\eta, 1+\eta]^3 \times \mathbb{I}}} |\frac{\partial \phi}{\partial a}(x,y,z,t)|, 1),$$

and let $\nu > 0$ be chosen so that the ν-neighborhood of $\mathbb{I}^2 \times \{\frac{1}{2}\}$ in the square metric is contained in M.

Consider the homotopy given by

$$H((x,y,z),t) =$$
$$(\phi(x,y,z,t)a(x,y) + x,$$
$$\phi(x,y,z,t)b(x,y) + y, \ \phi(x,y,z,t)c(x,y) + z)$$

For fixed t, we have

$$T(H_t) = \begin{bmatrix} \phi_x a + \phi a_x + 1 & \phi_x b + \phi b_x & \phi_x c + \phi c_x \\ \phi_y a + \phi a_y & \phi_y b + \phi b_y + 1 & \phi_y c + \phi c_y \\ \phi_z a + \phi a_z & \phi_z b + \phi b_z & \phi_z c + \phi c_z + 1 \end{bmatrix}.$$

It is an easy exercise to show that if r is the real root of

$$6x^3 + 6x^2 + 3x - 1,$$

then any real matrix whose entries each differ from the identity matrix by less than r is invertible. (It happens that $\frac{1}{4} > r > \frac{1}{5}$.)

It therefore follows that if we choose ϵ to be $\min(\frac{r}{2\mu}, \nu)$, then H_t is a diffeomorphism for all t (by the inverse function theorem). But $H_0 = Id_{[-\eta, 1+\eta]}$; $H((x,y,\frac{1}{2}),1) = \psi(x,y)$; and $H|N \times \mathbb{I}$ is trivial. $\square$

To adjust the tangents along the source and target, we construct ambient isotopies as follows: choose disjoint neighborhoods of each component of the source (resp. target) in $\mathbb{I}^3$, and coördinates on each neighborhood so that the following conditions are satisfied:

- the embedding of the source (resp. target) component has the identity map as z-coördinate

- its extension to an embedding of $(-\epsilon, 1 + \epsilon)$ also lies linearly in the z direction

- the yz-plane lies in the top (resp. bottom) face of $\mathbb{I}^3$

- the intersection of the spine of the ribbon tangle lies along the x-axis in such a way that the neighborhood (or a smaller one) is identified with $(-\epsilon, 1 + \epsilon) \times \mathbb{R} \times [0, \epsilon)$.

Then we apply the ambient isotopy of the following lemma to the neighborhood of each boundary component:

Lemma 8.11 *Let* $b : [-\epsilon, 1 + \epsilon] \times [0, \epsilon] \to \mathbb{R}^3$ *such that* $b(z, 0) = (0, 0, z)$ *and* $b(0, x) = (x, 0, 0)$, *and the image of* b *lies entirely in the closed positive or negative half-space with respect to the* z-*coördinate. Assume w.l.o.g. that the image lies in the positive half-space. Then there is an isotopy* H_t *of the half-space which fixes the complement of* $[0, \frac{\epsilon}{2}) \times (-N, N) \times (-\frac{\epsilon}{2}, 1 + \frac{\epsilon}{2})$, *fixes the boundary of the half-space, and such that if* $\vec{v}$ *is a unit vector normal to the* z-*axis and tangent to the image of* b, *then* $T(H_1)(\vec{v}) = (1, 0, 0)$.

proof: Now, let $m(z)$ be the slope of the vector normal to the z-axis and tangent to the image of b at $(0, 0, z)$. Let

$$
\mu(x, y, z, t) =
$$
$$
B(0, 1, -\frac{\epsilon}{2}, 1 + \frac{\epsilon}{2}, 1, 0, z) B(-1, 1, -N, N, 1, 0, y)
$$
$$
\Phi(\frac{\epsilon}{4}, 1, \frac{\epsilon}{2}, 0, x) \Phi(0, 0, 1, 1, t) m(z),
$$

where N is to be chosen later.

Define an isotopy by

$$
H((x, y, z), t) = (\frac{x}{\sqrt{1 + \mu(x, y, z, t)^2}}, y + \frac{\mu(x, y, z, t)x}{\sqrt{1 + \mu(x, y, z, t)^2}}, z)
$$

Now, observe that for any fixed t we have

$$T(H_t) =$$

$$
\begin{bmatrix}
\dfrac{1}{(1+\mu^2)^{\frac{1}{2}}} - \dfrac{x\mu\mu_x}{(1+\mu^2)^{\frac{3}{2}}} & -\dfrac{x\mu\mu_y}{(1+\mu^2)^{\frac{3}{2}}} & -\dfrac{x\mu\mu_z}{(1+\mu^2)^{\frac{3}{2}}} \\[4mm]
\dfrac{\mu_x x}{(1+\mu^2)^{\frac{1}{2}}} + \dfrac{\mu}{(1+\mu^2)^{\frac{1}{2}}} - \dfrac{x\mu^2\mu_x}{(1+\mu^2)^{\frac{3}{2}}} & 1 + \dfrac{\mu_y x}{(1+\mu^2)^{\frac{1}{2}}} - \dfrac{x\mu^2\mu_y}{(1+\mu^2)^{\frac{3}{2}}} & \dfrac{\mu_z x}{(1+\mu^2)^{\frac{1}{2}}} - \dfrac{x\mu^2\mu_z}{(1+\mu^2)^{\frac{3}{2}}} \\[4mm]
0 & 0 & 1
\end{bmatrix}
$$

This is invertible if and only if the xy-minor is, but the determinant of this minor is given by

$$
\frac{1}{(1+\mu^2)^{\frac{1}{2}}} + \mu_y \frac{x(1+\mu^2) - x^2\mu\mu_x}{(1+\mu^2)^2}.
$$

By construction, μ is bounded above and below by

$$
\pm M = \pm \max_{z\in[-\epsilon,1+\epsilon]} |m(z)|,
$$

$1 + \mu^2$ is bounded below by 0, and μ_x is bounded above and below by plus and minus

$$
\max_{x\in[(\frac{\epsilon}{4},\frac{\epsilon}{2}]} |\Phi'(\frac{\epsilon}{4}, 1, \frac{\epsilon}{2}, 0, x)| M
$$

respectively, for any choice of N. Now, we can choose N so that the maximum value of $|\mu_y|$ is arbitrarily small. Thus, by choosing N sufficiently large, we can make the determinant arbitrarily close to $\frac{1}{(1+\mu^2)^{\frac{1}{2}}}$, which is bounded away from zero by $\frac{1}{(1+M^2)^{\frac{1}{2}}}$. Thus, by the inverse function theorem, each H_t is a diffeomorphism.

The fact that the isotopy fixes the desired regions is immediate by construction. $\square$

After applying the isotopies provided by this lemma to straighten the 1-jet neighborhood of the source and target of one ribbon tangle to match the other, we wish to apply an ambient isotopy which will shrink the ribbon along itself until it lies in a very small neighborhood of the union of the source, target and spine in which the ambient isotopy provided by Lemma 8.10 will complete the proof.

To construct the desired ambient isotopy and complete the proof of the approximation theorems used in the proof of Theorems 8.5 and 8.6, we prove:

Lemma 8.12 *Let $0 < \epsilon \leq 1$, and let X be any smooth manifold and $b : X \to \mathbb{I}$ be any smooth function bounded below by ϵ. Then there exists an isotopy $h_b : X \times [-\epsilon, 1+\epsilon] \times \mathbb{I} \to X \times [-\epsilon, 1+\epsilon]$ from the identity map to a map $s_b(x, z)$ satisfying*

- $s_b(x, 0) = (x, 0)$

- $s_b(x, 1) = (x, b(x))$

- $s_b(x, z) = (x, \zeta)$ *for some* ζ

- $\frac{\partial s_b}{\partial z}(x, z) = 1$ *for all x and all* $z \in [-\epsilon, \frac{\epsilon}{6}] \cup [1-\frac{\epsilon}{6}, 1+\frac{\epsilon}{6}] \cup [1+\frac{5\epsilon}{6}, 1+\epsilon]$

- $\frac{\partial^n s_b}{\partial z^n}(x, z) = 0$ *for all $n \geq 2$, all x and all*
  $z \in [-\epsilon, \frac{\epsilon}{6}] \cup [1 - \frac{\epsilon}{6}, 1+\frac{\epsilon}{6}] \cup [1+\frac{5\epsilon}{6}, 1+\epsilon]$,

and $\exists \delta > 0$ such that

- $h_b(x, p, t) = (x, p)$ *for all $t \in [0, \delta)$*

- $h_b(x, p, t) = s_b(x, p)$ *for all $\in (1 - \delta, 1]$.*

proof: Let β be a smooth bump function for $[\frac{\epsilon}{3}, 1-\frac{\epsilon}{3}] \subset (\frac{\epsilon}{6}, 1-\frac{\epsilon}{6})$ with inner value 0 and outer value 1.

Then

$$\sigma_a(r) = \int_0^r 1 - a\beta(p)dp$$

for $a \in [0, 1)$ is a smooth monotone increasing function which satisfies

- $\sigma_a(0) = 0$

- $\sigma_a'|_{(-\infty,\frac{\epsilon}{6}]\cup[1-\frac{\epsilon}{6},\infty)} = 1$

- $\sigma_a^{(n)}|_{(-\infty,\frac{\epsilon}{6}]\cup[1-\frac{\epsilon}{6},\infty)} = 0$ for $n \geq 2$.

Moreover, if we let $\xi(a) = \sigma_a(1)$, then $\xi(0) = 1$; ξ is smooth and strictly decreasing; and $\lim_{a\to 1}\xi(a) < \frac{2\epsilon}{3}$, since $1 - \phi(p) \leq f(p)$, where

$$f(p) = \begin{cases} 1 & \text{if } p \in [0,\frac{\epsilon}{3}) \cup (1 - \frac{\epsilon}{3}, 1] \\ 0 & \text{if } p \in [\frac{\epsilon}{3}, 1 - \frac{\epsilon}{3}] \end{cases}.$$

Now, let $a(\xi)$ denote the inverse function, and let

$$\psi_b(x,p) = \Phi(\delta, 0, 1 - \delta, a(b(x), p)).$$

Then $\eta_{b,[0,1]}(x,p,t) = (x, \sigma_{\psi_b(x,t)}(p))$ is the restriction of the desired h_b to $X \times [0,1]$.

A similar construction using a smooth bump function θ for

$$\left[1 + \frac{\epsilon}{3}, 1 + \frac{2\epsilon}{3}\right] \subset \left(1 + \frac{\epsilon}{6}, 1 + \frac{5\epsilon}{6}\right)$$

with inner value 1 and outer value 0, and $\tau_a(r) = 1 + \int_1^r 1 + a\theta(p)dp$ gives $\eta_{b,[1,1+\epsilon]}(x,p,t)$, with the necessary properties to be the restriction of h_b to $X \times [1, 1+\epsilon]$.

We then have

$$h_b(x,p,t) = \begin{cases} (x,p) & \text{if } p \in [-\epsilon, 0) \\ \eta_{b,[0,1]}(x,p,t) & \text{if } p \in [0,1] \\ \eta_{b,[1,1+\epsilon]}(x,p,t) & \text{if } p \in [1, 1+\epsilon] \end{cases}.$$

□

Finally,

Lemma 8.13 *If D is a disk (ball) of any dimension of radius r, the isotopy of the previous proposition extends to an ambient isotopy H_b of $X \times [-\epsilon, 1+\epsilon] \times D$ whose restriction to $X \times [-\epsilon, 1+\epsilon] \times \{0\}$ is h_b, and which is trivial on a neighborhood of the boundary of D.*

proof: Let $H_b(x, p, v, t) = h_b(x, p, t\phi(1 - \frac{|v|^2}{r^2}), v)$, where ϕ is a smooth ramp function from $(\epsilon, 0)$ to $(1 - \epsilon, 1)$ for some $0 < \epsilon < \frac{1}{2}$. $\square$

This, in fact, completes the proof of Lemma 8.9 and with it the proof of Theorem 8.6: as indicated above, the ambient isotopy of Lemma 8.9 is obtained by composing (in the sense of isotopies) a sequence of ambient isotopies. First, we apply those obtained from Lemma 8.11 for each component of the source and target by trivial extension to $\mathbb{I}^3$ outside of the neighborhoods on which they are defined. Then, we apply the ambient isotopy of Lemmas 8.13 and applied to a tubular neighborhood of the ribbon tangle for a smooth function chosen so that its graph lies in a neighborhood of the source, spine and target of the ribbon tangle small enough that the bounds on the embedding and its partial derivatives in the hypotheses of Lemma 8.10 apply in a chosen coördinate system, and finally we apply the inverse of the corresponding ambient isotopy for the other ribbon tangle. $\square$

Thus, the abuse of language of speaking of "the category of tangles" or "the category of framed tangles" without specifying how the category in question was constructed from topological data is justified to the extent that all such usages are: up to a structure preserving equivalence of categories.

Chapter 9

Shum's Theorem

One of the most remarkable theorems proven in the past two decades is Mei-Chi Shum's coherence theorem for ribbon categories [48, 49]. For those whose sense of categorical coherence theorems is fixed on the classical "all diagrams commute" theorems of Mac Lane [39] and Epstein [21], which simply make trivial things that ought to be trivial, it may seem strange that a categorical coherence theorem should be of great importance.

It has, however, long been known in the Australian school of category theory that the correct general notion of a coherence theorem is a characterization up to equivalence of a category with certain structure freely generated by a given category (or directed graph). That being said, we can state the simplest instance of Shum's theorem:

Theorem 9.1 *The ribbon category freely generated by a single object is monoidally equivalent to* **FrTang**.

When it is remembered that the categories of representations of quantized universal enveloping algebras (among other categories) are all ribbon categories, this theorem provides the explanation of the remarkable connection between Hopf algebra theory and knot theory: given any ribbon category $\mathcal{C}$ and an object X therein as an image of the generating object (say the category of representations of a quasi-triangular Hopf

117

algebra, and a particular representation), the freeness of **FrTang** induces a ribbon functor (monoidal functor preserving the braiding, twist and duality) $\Phi_X : \textbf{FrTang} \to C$. In cases in which $Hom_C(I, I)$ is a ring R (as, for example, categories of representations of a quasi-triangular Hopf algebra), Φ_X then induces an R-valued invariant of framed links.

All of those values of the HOMFLY [42] and Kauffman [32] polynomials which "come from quantum groups" are examples of this type of "functorial invariant." Specifically, the values of the HOMFLY polynomial in the normalization given in Chapter 6 with $x = q^{-n}$ and $z = q^{\frac{1}{2}} - q^{-\frac{1}{2}}$ arise as functorial invariants by mapping the downward oriented strand to the fundamental representation of $U_q(sl_n)$. Similarly, there are values of the Kauffman polynomial corresponding to the fundamental represenyations of quantized universal enveloping algebras for simple Lie algebras of types B, C and D.

Rather than prove Theorem 9.1 directly, we will instead set up the machinery to prove the full version of Shum's coherence theorem. In general we follow [49], but with some differences necessitated by our more geometric definition of tangles:

Definition 9.2 *A framed tangle labeled by a category C is a framed oriented tangle, each boundary point of which is labeled with an object of C, and each component of which is labeled with a map of C. subject to the restrictions:*

1. *If a component is not a closed curve, then the source (resp. target) of its labeling map is the labeling of its first (resp. second) endpoint relative to the orientation.*

2. *If a component is a closed curve, then its labeling map is an element of $E(C) = \coprod_{A \in Ob(C)} C(A, A)$.*

Definition 9.3 *Two framed tangles T, T' labeled by C are* equivalent *if there is a 1-1 correspondence between the components of T and those of T' such that the following hold:*[1]

---

[1]There is a slight inadequacy in the corresponding definition in [49]: it is insuffi-

1. *The underlying framed oriented tangles of T and T' are equivalent via an ambient isotopy rel boundary $H : \mathbb{I}^3 \times \mathbb{I} \to \mathbb{I}^3$ such that if $H(S, 0)$ is a component of T then $H(S, 1)$ is the corresponding component of T'.*

2. *If c is a component of T which is not a closed curve and c' is the corresponding component of T', then the labeling maps (resp. the labeling object of the first endpoint with respect to the orientation, the labeling object of the second endpoint with respect to the orientation) of c and c' are equal.*

3. *If c is a closed component of T and c' is the corresponding component of T', then the labeling maps are equivalent with respect to the equivalence relation $\equiv$ "trace equivalence" on $E(\mathcal{C})$ induced by*

$$fg \equiv gf.$$

Definition 9.4   *Given a category $\mathcal{C}$, the category of $\mathcal{C}$-labeled framed tangles, denoted* **FrTang** $\int \mathcal{C}$*, has as objects finite framed sets of points in $(0, 1)^2$, each point of which is equipped with an orientation (i.e. a sign) and a label by an object of $\mathcal{C}$. As for tangles, a framing is specified by extending the embedding of the finite set S to an embedding of $S \times \mathbb{I}$.*

The arrows of **FrTang** $\int \mathcal{C}$ *are equivalence classes of framed tangles labeled by $\mathcal{C}$, with source (resp. target) given by the intersection with the face $\mathbb{I}^2 \times \{0\}$ (resp. $\mathbb{I}^2 \times \{1\}$) with the induced orientation, framing and labeling by objects of $\mathcal{C}$. Identity maps are identity framed tangles labeled with identity maps from $\mathcal{C}$. Composition is given by composing the underlying framed tangles and composing the labeling maps.*

Observe that the restrictions relating the labelings on the source and target ensure that the labels of the composition on non-closed components are well-defined, while the labels of the composition on closed components are well-defined up to trace equivalence.

---

cient to specify that the underlying framed tangles (or double tangles in Shum's terminology) are equivalent – one must specify the correspondence between the components as well. To see why, consider the case of a tangle consisting of a 0-framed unlink of two components with the components labeled by trace-inequivalent endomorphisms.

We then have

Proposition 9.5 *For any category C, the category* **FrTang** $\int C$ *is a ribbon category.*

proof: Most of the work has already been done in the proof of Proposition 7.4. The structure maps for **FrTang** $\int C$ are all given by labeling with identity maps all components of a framed tangle representing the corresponding structure map of **FrTang**. All of the coherence conditions follow from the corresponding condition in **FrTang**, so that all that remains is to check naturality for α, σ, ρ, λ and θ. In all cases, this follows from the same argument as for **FrTang**, together with an application of the rather trivial observation that if $f : X \to Y$ is a map in C, then we can use the equation $Id_X f = f Id_Y$ to move labeling maps past the identity label in the structure map. □

With this we can now state Shum's coherence theorem [48, 49]:

Theorem 9.6 *Let $F(C)$ be the free ribbon category generated by the category C. Then the functor $\Phi : F(C) \to$ **FrTang** $\int C$ induced by the freeness condition is an equivalence of ribbon categories, that is, a monoidal equivalence, both functors of which preserve the braiding, twist, dual objects and structure maps for dual objects.*

proof: This takes a bit of work. First we must see that our geometrically defined category of framed tangles is equivalent to a combinatorially defined version in terms of diagrams, and then, as in Shum [48, 49] show that the combinatorial diagrammatic category of tangles is equivalent to the syntactically constructed free ribbon category generated by C.

We prove a sequence of lemmas:

Lemma 9.7 *The category of C-labeled framed tangles* **FrTang** $\int C$ *is monoidally equivalent to its full subcategory* **FTC**, *whose objects have underlying point sets of the form $\{(s, \frac{1}{2}) | s \in S\}$ for some set S in the family $\mathcal{S}$ of finite subsets of $\mathbb{I}$ described inductively by*

- $\{\frac{1}{2}\} \in S$

- If $\frac{n}{2^k} \in S \in S$ for n odd, then $[S \setminus \frac{n}{2^k}] \cup \{\frac{2n-1}{2^{k+1}}, \frac{2n+1}{2^{k+1}}\} \in S$

- If $T \subset S \in S$, then $T \in S$

and whose framing ribbons lie to the right along the line $\{(x, \frac{1}{2})\}$ provided
the orientation at the point is positive, and to the left along the same
line provided the orientation at the point is negative, and are of width
$\frac{1}{2^{k+2}}$ when the point's x-coördinate is of the form $\frac{2n-1}{2^k}$.

Moreover, the inclusion is a strict ribbon functor, that is, a strict
monoidal functor which strictly preserves the braiding, balancing, dual
objects and structure maps for duals.

proof: It is immediate that **FTC** is itself a ribbon category once it
is observed that it is closed under $\otimes$ and $(-)^*$, and that the obvious
inclusion functor, which we denote ι, is a strict ribbon functor.

As is often the case when a full subcategory is shown to be equivalent
to its ambient category, we construct the retraction functor R along with
the natural isomorphisms which implement the equivalence. Heuristi-
cally, we need to choose for each object of **FrTang** $\int C$ an object of the
full subcategory with the same number of underlying points of each sign
and object label, and a (framed) geometric tangle (each strand of which
is labeled with an identity map from C) joining the given object with
the chosen object of the subcategory. With care, we can do this so that
whenever the object is already in the subcategory, the identity framed
geometric tangle will be chosen. Specifically, we proceed as follows:

Lexicographically order the points of $\mathbb{I}^2$ with the x-coördinate dom-
inant; that is, let $(a, b) < (c, d)$ whenever $a < c$ or both $a = c$ and
$b < d$. Likewise, for each cardinality of set in S consider each set as
a monotonically increasing word of elements of $\mathbb{I}$ and lexicographically
order them.

Now, for each object X of **FrTang** $\int C$, construct an image object
$R(X)$ and an isomorphism from X to $\iota(R(X))$ as follows: Let $\hat{X}$ be the
underlying point-set of X. Consider an element S of S such that $S \times \{\frac{1}{2}\}$
is contained in $\hat{X}$, and of maximum cardinality among such. If there is

more than one, choose the earliest in the lexicographic order. Let $S' = S$ if S is non-empty or $\{\frac{1}{2}\}$ if it is empty. Now, let $S' = \{\xi_1, \ldots, \xi_r\}$ with $\xi_1 < \xi_2 < \ldots < \xi_r$ and let $T = \{x_0, \ldots x_r\}$, where $x_0 = 0$ and $x_r = 1$ and $x_i = \frac{\xi_i + \xi_{i+1}}{2}$. Let

$$
y_i = \begin{cases} 0 & \text{if } i = 0 \\ 1 & \text{if } i = 1 \\ \frac{1}{2} & \text{otherwise} \end{cases}.
$$

Now, let $\hat{X}_i = \{(x, y) \in \hat{X} | (x_i, y_i) < (x, y) < (x_{i+1}, y_{i+1})\}$. Let S'' denote the set obtained from S' by iteratively replacing the smallest element of S' (or its succeeding sets) lying in $\{(x, y) | (x_i, y_i) < (x, y) < (x_{i+1}, y_{i+1})\}$ with two elements, as in the second condition defining S, until $S''_i = \{(x, y) \in S'' | (x_i, y_i) < (x, y) < (x_{i+1}, y_{i+1})\}$ is of equal cardinality to $\hat{X}_i$.

$R(X)$ is then the object of **FTC** with underlying point set given by $\widehat{R(X)} = \{(x, \frac{1}{2}) | x \in S''\}$, with labels and orientation given so that the order of the labels and orientations is the same in X and $R(X)$, where in each case we induce an ordering using the lexicographic order on $\mathbb{I}^2$. Then there is a geometric braid consisting of straight line segments joining $\hat{X} \times \{0\}$ and $\widehat{R(X)} \times \{1\}$. This braid is the underlying tangle of a C-labeled framed tangle with identity maps for all labels and framing ribbons which rotate clockwise by less than a full turn. This map, then, is an isomorphism $\phi_X : X \to \iota(R(X))$.

The construction of $R(X)$, then, is extended to a functor defining $R(f) : R(X) \to R(Y)$ for $f : X \to Y \in \textbf{FrTang} \int C$ by $R(f) = \phi_X^{-1} f \phi_Y$ (observe that the composite of maps in **FrTang** $\int C$ is in the full subcategory *by fullness*).

Now, by construction $R(X \otimes Y) = R(X) \otimes R(Y)$ and $R(\iota(X)) = X$, and similarly for maps in **FTC**; and ϕ is a monoidal natural isomorphism from X to $\iota(R(X))$.

It follows from the naturality of σ in each variable and from the monoidal naturality of ϕ, that $R(\sigma_{A,B}) = \sigma_{R(A),R(B)}$. Similarly, it follows that $R(A^*)$ is a right-dual of $R(A)$ (with structure maps given by $R(\eta)$ and $R(\epsilon)$) and that $R(A^*)$ is thus canonically isomorphic to $R(A)^*$. □

In a sense, the previous lemma is half of the story. The other half is given by the reduction of the syntactically constructed free ribbon category on $\mathcal{C}$ to a monoidally equivalent full subcategory which will be easily seen to be isomorphic(!) to **FTC**.

We recall from Shum [48, 49] the syntactical construction of the $F(\mathcal{C})$: its objects are the free $(I, \otimes, (-)^*)$ algebra on $Ob(\mathcal{C})$. The arrows are described by specifying a graph of generating arrows, freely generating a category and quotienting by all instances of the relevant relations, specifically, for the graph $H(\mathcal{C})$ whose vertices are $Ob(F(\mathcal{C}))$, with directed edges (arrows) of the following types:

- For each object X, arrows of the following types (with sources and targets as indicated):

  ◇ $\rho_X : X \otimes I \to X$
  ◇ $\bar{\rho}_X : X \to X \otimes I$
  ◇ $\lambda_X : I \otimes X \to X$
  ◇ $\bar{\lambda}_X : X \to I \otimes X$
  ◇ $\theta_X : X \to X$
  ◇ $\bar{\theta}_X : X \to X$
  ◇ $\eta_X : I \to X^* \otimes X$
  ◇ $\epsilon_X : X \otimes X^* \to I$

- For each pair of objects X, Y, arrows of the following types:

  ◇ $\sigma_{X,Y} : X \otimes Y \to Y \otimes X$
  ◇ $\bar{\sigma}_{X,Y} : Y \otimes X \to X \otimes Y$

- For each triple of objects X, Y, Z, arrows of the following types:

  ◇ $\alpha_{X,Y,Z} : [X \otimes Y] \otimes Z \to X \otimes [Y \otimes Z]$
  ◇ $\bar{\alpha}_{X,Y,Z} : X \otimes [Y \otimes Z] \to [X \otimes Y] \otimes Z$

- For each arrow $f : A \to B$ in $\mathcal{C}$ an edge $[f] : A \to B$ (where A and B are regarded as the objects in $F(\mathcal{C})$)

- For each edge $e : W \to X \in H(\mathcal{C})$ and each object $Z \in Ob(F(\mathcal{C}))$ edges of the following types:

  ⋄ $Z \otimes e : Z \otimes W \to Z \otimes X$

  ⋄ $e \otimes Z : W \otimes Z \to X \otimes Z.$

Given an arrow $t \in H(\mathcal{C})$, an *expansion* of t is t or any arrow formed from t and objects of $F(\mathcal{C})$ by iterated application of the last construction. Let $K(\mathcal{C})$ denote the category freely generated by $H(\mathcal{C})$, that is the category whose objects the vertices of $H(\mathcal{C})$, whose arrows are all paths along edges in $H(\mathcal{C})$, including as paths of length zero at each vertex as identity arrows, and whose composition is given by concatenation of paths. Relations then impose equivalences between these paths. By an *expansion* of a relation, we mean a relation obtained from the given one by replacing each map with a map obtained from the given map and objects by iterated application of the last construction (using the same objects and iteration scheme for each map in the relation).

$F(\mathcal{C})$ is then the quotient of $K(\mathcal{C})$ by all relations of the following forms and all their expansions:

- relations which make $\otimes$ into a functor:

$$(t \otimes W)(X \otimes s) \equiv (Y \otimes s)(t \otimes Z) : X \otimes Z \to Y \otimes W$$

$$Id_X \otimes Y \equiv X \otimes Id_Y \equiv Id_{X \otimes Y}$$

- relations which impose naturality of $\alpha, \overline{\alpha}, \rho, \overline{\rho}, \lambda, \overline{\lambda}, \sigma, \overline{\sigma}, \theta, \overline{\theta}$

- the pentagon and triangle relations for the monoidal structure

- the two hexagon relations for the braiding

- the defining relations for the balancing, and that relating the balancing to the duality

- relations imposing the conditions that x and $\bar{x}$ are inverses for $x = \alpha, \rho, \lambda, \sigma, \theta$

- the defining relations for right duals on η and ϵ

- relations to make the inclusion $\mathcal{C} \to \mathcal{F}(\mathcal{C})$ into a functor, namely

$$[f][g] \equiv [fg]$$

and

$$[Id_A] \equiv Id_A.$$

where the latter Id_A is the length zero path at the object A.

As Shum observes, the following result is obvious:

Proposition 9.8 *Every map in $F(\mathcal{C})$ is a composition of expansions of* $\alpha, \alpha^{-1}, \rho, \rho^{-1}, \lambda, \lambda^{-1}, \sigma, \sigma^{-1}, \theta, \theta^{-1}, \eta, \epsilon$ *and images of maps in* $\mathcal{C}$.

The balance of the proof of Shum's coherence theorem consists of finding a full subcategory of $F(\mathcal{C})$ which is monoidally equivalent to $F(\mathcal{C})$, and is isomorphic as a ribbon category to **FTC**:

Shum [49] follows Kelly and Laplaza [34] to introduce a simple notion of prime factorization in $F(\mathcal{C})$: An object is *prime* if it is an object of $\mathcal{C}$ or is of the form Y^* for some object Y. It is then clear that any object of $F(\mathcal{C})$ can be expressed uniquely as an iterated monoidal product of prime objects (considering I as the empty monoidal product).

An object of $F(\mathcal{C})$ is *reduced* if all of its prime factors are of the form A or A^* for $A \in \mathcal{C}$. A map is reduced if it is a composition of expanded instances of $\alpha, \alpha^{-1}, \rho, \rho^{-1}, \lambda, \lambda^{-1}, \sigma, \sigma^{-1}, \theta, \theta^{-1}, \eta, \epsilon$ and images of maps in $\mathcal{C}$ with reduced sources and targets.

Lemma 9.9 *The full subcategory RFC of reduced objects in $F(\mathcal{C})$ is monoidally equivalent to $F(\mathcal{C})$.*

proof: As always in showing that a full subcategory is equivalent to its ambient category, we proceed by constructing the image object and the isomorphism together. In this case, our objects are defined syntactically as elements of the free $(\otimes, I, (-)^*)$ algebra on $Ob(C)$. We define the isomorphism and its target $\phi : X \to R(X)$ by composing maps obtained recursively as follows:

1. If $X = Y \otimes Z$, then $\phi_X = [\phi_Y \otimes Z][R(Y) \otimes \phi_Z] = \phi_Y \otimes \phi_Z$

2. If $X = (Y \otimes Z)^*$, then $\phi_X = u_{Y,Z} \phi_{Z^* \otimes Y^*}$

3. If $X = Y^{**}$, then $\phi_X = c_Y \phi_Y$

4. If $X = I^*$, then ϕ_X is the canonical isomorphism from I^* to I

5. If $X = I$ or $X \in Ob(C)$, then $\phi_X = Id_X$

where $u_{Y,Z}$ is the canonical isomorphism from $(Y \otimes Z)^*$ to $Z^* \otimes Y^*$ induced by the fact that both are right duals to $Y \otimes Z$, and c_Y is the isomorphism from Y^{**} to Y given by the sovereign structure of the ribbon category, by considering Y^{**} as the double right dual.

The target of the composite map just defined is plainly reduced, and $R(X \otimes Y) = R(X) \otimes R(Y)$, $\phi_X = \phi_Y \otimes \phi_Z$, and if X is reduced then $R(X) = X$ and $\phi_X = Id_X$, all by construction. Thus, the full subcategory RFC is monoidally equivalent to $F(C)$. $\square$

More than that, we have

Lemma 9.10 *Every map in RFC is reduced.*

proof: By Proposition 9.8, any map can be written as a composition of expanded instances of structure maps and of images of maps in C. If the intermediate objects in the composition were reduced, we would be done. However, this need not be the case.

The trick is iteratively to replace the composition of expanded instances with others in such a way that the intermediate objects are guaranteed to be "closer" to being reduced. We follow Shum's method of measuring non-reduction with a syntactic "rank" k for objects in $F(\mathcal{C})$:

$$
\begin{aligned}
k(A) &= 0 \text{ for } A \in Ob(\mathcal{C}) \\
k(A^*) &= 0 \text{ for } A \in Ob(\mathcal{C}) \\
k(I) &= 0 \\
k(I^*) &= 1 \\
k([X \otimes Y]^*) &= k(Y^*) + k(X^*) + 1 \\
k(X^{**}) &= 3k(X^*) + 1 \\
k(X \otimes Y) &= k(X) + k(Y).
\end{aligned}
$$

It is almost immediate by construction that $k(RX) < k(X)$ whenever X is not already reduced: it suffices to observe that the initial factor in each of items 2, 3 and 4 in the definition of ϕ are strictly rank reducing, and that a monoidal product is reduced if and only if its factors are reduced. From this, it follows that any non-reduced object is replaced by an object of lower rank, since one of items 2, 3, or 4 must apply to some monoidal factor.

The replacement step then is given by the observation that any (expanded) instance of a structure map $\tau : X \to Y$ may be replaced by a composition of the form

$$
\phi_X s_1 \dots s_n \phi_Y^{-1}
$$

where each of the intermediate objects in the composition has syntactic rank less than or equal to the larger of $k(R(X))$ and $k(R(Y))$. As Shum observes, the fact that rank is additive under $\otimes$ and that R is a monoidal functor implies that it suffices to show this result for non-expanded instances of the structure maps. This is left as an exercise for the reader (the only non-trivial cases are instances of η and ϵ, where the syntactic form of the object to which they are applied must be considered), or the reader may refer to the proof of Lemma 3.2 in [49]. $\square$

To prove Shum's coherence theorem, it thus suffices to show

Lemma 9.11 *RFC and* **FTC** *are isomorphic as ribbon categories.*

proof: To begin, we construct a bijection between the sets of objects: First construct a bijection between the sets in S and the elements of the free $(\otimes, I)$-algebra $F_{\otimes, I}(g)$ on one generator g by using the well-known bijection between binary trees (describing the step of replacing one point of an element of S with two to obtain a new element of S) and parenthesization schemes, but matching an occurrence of I in a leaf to the removal of a point from the set of S. (Thus, for example, $\{\frac{1}{4}, \frac{7}{8}\}$ corresponds to $g \otimes (I \otimes g)$.)

Now, every object of RFC is given by an element of the free $(\otimes, I)$-algebra generated by $DlC = Ob(\mathcal{C}) \cup \{A^* | A \in Ob(\mathcal{C})\}$, and thus may be regarded as the labeling of the g's in an underlying element of $F_{\otimes, I}(g)$ with elements of DlC.

By replacing $A \in Ob(\mathcal{C})$ with $(A, +)$ and A^* for $A \in Ob(\mathcal{C})$ with $(A, -)$, and using these labels on the corresponding points of the element of S corresponding under the bijections described above, we obtain a corresponding object of **FTC**. It is clear that the map from $Ob(RFC)$ to $Ob(\mathbf{FTC})$ just constructed is bijective and preserves both $\otimes$ and I.

On maps, we now use Lemma 9.10 to write any map in RFC as a composition of expanded instances of structure maps applied to prime objects. Each factor has as source and target a reduced object, which has an image in **FTC** constructed above with the same syntactical structure in terms of the generating objects, their duals, $\otimes$ and I as the corresponding object in RFC. Thus these are the source and target of an expanded instance of the corresponding structure maps in **FTC**. Therefore, we have a factorization of functors as shown in Figure 9.1.

It remains to show that the factorization functor just constructed is an equivalence of ribbon categories. Observe that the preservation properties are immediate from the preservation properties of the inclusion of RFC and the universal functor from the free ribbon category. It remains only to show

Lemma 9.12 *Every equation of maps in* **FTC** *follows from the relations of ribbon categories.*

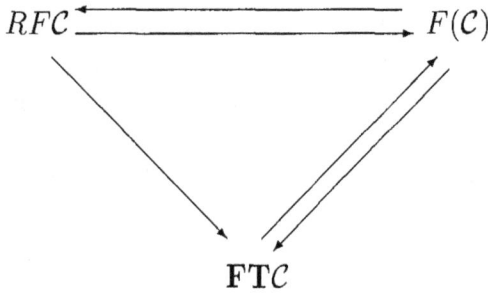

Figure 9.1: The Functor from RFC to $F(\mathcal{C})$ Induced by Inclusion and Freeness Factors through **FT$\mathcal{C}$**

proof: Once it is observed that relations which commute maxima, minima and crossings (that is, which modify the relative height of features of the tangle without changing the diagram) follow from the functoriality of $\otimes$, it suffices to see that the relations given by the framed Reidemeister moves are induced by the relations of ribbon categories.

$\Delta.\pi.1$ is the duality relation on generating objects and their duals. $\Delta.\pi.2$ follows from the invertibility of σ and its naturality. $\Omega.2$ with the strands running vertically is simply the invertibility of σ, while $\Omega.2$ with the strands running horizontally follows from the vertical version and use of $\Delta.\pi.1$ and $\Delta.\pi.2$. $\Omega.3$ follows from the hexagon conditions on the σ and the naturality of σ and σ^{-1}.

The only really troublesome relation is $\Omega.1_f$.

For $\Omega.1_f$ it is necessary to consider carefully what exactly is encoded by the maxima and minima with each orientation. A maximum with the left strand oriented up (resp. down) is an instance of η (resp. h), the structure map for a right (resp. left) dual. A minimum with the left strand oriented down (resp. up) is an instance of ϵ (resp. e), the other structure map for a right (resp. left) dual.

But e_X and h_X are given in terms of the other structure maps of a ribbon category by $e_X = [X \otimes \theta_{X^*}^{-1}]\sigma_{X,X^*}\epsilon_X$ and $h_x = \eta_X \sigma_{X^*,X}^{-1}[\theta_X \otimes X^*]$, respectively.

It then follows by composing with $\theta_I^{-1} \otimes X = Id_{I \otimes X}$, applying the naturality of θ^{-1} and the (inverse of) the coherence condition relating θ and σ, and the combination of naturality and dinaturality which simulates the "Whitney trick", that the strand with a positive crossing loop on the left is equal to θ^{-1}. That the strand with a positive crossing loop on the right has the same value follows from the "Whitney trick" alone. □ □

Thus, we have established Shum's coherence theorem. □

Beyond explaining the connection between knot theory and the representations of quantum groups, this theorem also justifies the use of a very convenient notation for maps in ribbon categories: maps may be depicted as framed tangle diagrams with strands labeled by objects, with the added feature of points on the strings (possibly including singularities where serval strands meet) labeled by specific maps not generated by the structure maps. The topology of the tangle diagram then takes care of all of the equations which follow from the coherence conditions, while other equations can be handled by cutting out a part of the diagram giving one representation of a map, and replacing it with another.

This sort of notation was first used explicitly as a computational tool by the author in [60], though it was developed independently at about the same time by Reshetikhin and Turaev [46] for use in the description of topological invariants.

Chapter 10

A Little Enriched Category Theory

The observation in Chapter 2, that given two categories, there is, in fact, a category of functors between them, rather than just a set of functors between them, suggests the possibility of a more general notion than that of category, in which one begins with a "suitable" category, and has "hom-objects" taken from that category.

It turns out that the right notion of "suitable" has already been given: a symmetric monoidal category. In the exposition below we follow, in general, the classic book by Kelly [33], to which the reader is referred for a more extensive treatment.

Definition 10.1 *Let $(\mathcal{V}, \otimes, I, \alpha, \rho, \lambda, \sigma)$ be a symmetric monoidal category. A $\mathcal{V}$ enriched category or simply $\mathcal{V}$-category $\mathcal{X}$ is a collection of objects $Ob(\mathcal{X})$ and an assignment to each pair of objects A, B of an object of $\mathcal{V}$, $\mathcal{X}(A, B)$, together with maps in $\mathcal{V}$*

$$Id_A : I \to \mathcal{X}(A, A)$$

for each object A, and

$$\circ = \circ_{A,B,C} : \mathcal{X}(A, B) \otimes \mathcal{X}(B, C) \to \mathcal{X}(A, C)$$

for each triple of objects A, B, C, satisfying

$$(\mathcal{X}(A,B) \otimes \mathcal{X}(B,C)) \otimes \mathcal{X}(C,D) \xrightarrow{\;\;a\;\;} \mathcal{X}(A,B) \otimes (\mathcal{X}(B,C) \otimes \mathcal{X}(C,D))$$

$\circ \otimes Id$ $\qquad\qquad\qquad\qquad\qquad\qquad\qquad\qquad$ $Id \otimes \circ$

$$\mathcal{X}(A,C) \otimes \mathcal{X}(C,D) \qquad\qquad\qquad\qquad \mathcal{X}(A,B) \otimes \mathcal{X}(B,D)$$

$$\circ \qquad\qquad \mathcal{X}(A,D) \qquad\qquad \circ$$

and

$$\mathcal{X}(A,A) \otimes \mathcal{X}(A,B) \xrightarrow{\;\;\circ\;\;} \mathcal{X}(A,B) \xleftarrow{\;\;\circ\;\;} \mathcal{X}(A,B) \otimes \mathcal{X}(B,B)$$

$Id_A \otimes Id$ $\qquad\qquad l \qquad\qquad\qquad r \qquad\qquad$ $Id \otimes Id_B$

$$I \otimes \mathcal{X}(A,B) \qquad\qquad\qquad\qquad\qquad \mathcal{X}(A,B) \otimes I$$

For the purposes of this study, the most important classes of examples have special names given in the definitions which follow:

Definition 10.2 *For R a commutative ring, an R-linear category is an R-**mod** enriched category, where R-**mod** has the monoidal structure given by $(\otimes_R,\ I = R, \ldots)$*

and

Definition 10.3 *For R a local commutative ring with maximal ideal* m, *a* complete R-linear category *is an* $\mathcal{X}_R$-*category, where* $\mathcal{X}_R$ *is the category of* m-*adically complete R-modules with monoidal structure given by* $(\widehat{\otimes}_R,\ I = R, \ldots)$, *where* $\widehat{\otimes}_R$ *is the* m-*adic completion of the algebraic tensor product of R-modules.*

We will also have call to consider

Definition 10.4 *A* topologized category *is a category enriched in* (**Esp**, $\times, \{*\}, \ldots$).[1]

Definition 10.5 *A* stratified space *is a space X equipped with a filtration* $X = X_0 \supset X_1 \supset X_2 \supset \ldots$. [2] *The set-difference* $X_i \setminus X_{i+1} = S_i$ *is called the* i^{th} stratum *of X. The* finite codimension part *of a stratified space X given by* $X^{\mathrm{fin}} = \cup_{i=0}^{\infty} X_i \setminus X_{i+1}$ *is a stratified space with filtration* $X_i^{\mathrm{fin}} = X_i \cap X^{\mathrm{fin}}$. *A* stratified map $f : X \to Y$ *is a continuous map which respects the filtration in the sense* $f|_{X_i}$ *factors through the inclusion of* Y_i. *Stratified spaces and stratified maps then form a monoidal category* **Strat** *when equipped with the product* $\otimes$ *given by letting* $X \otimes Y$ *be the stratified space with underlying topological space* $X \times Y$ *and strata* $(X \otimes Y)_i = \cup_{j=0}^{i} X_j \times Y_{i-j}$, *and with unit* $\mathbf{1} = \{*\} \supset \emptyset \supset \emptyset \supset \ldots$. *(Observe that the obvious associator and unit transformations inherited from* **Esp** *are stratified maps, and thus provide the necessary structure.)*

Definition 10.6 *A* stratified category *is a category enriched in* **Strat**.

Once one has enriched categories, one needs enriched functors:

Definition 10.7 *A* $\mathcal{V}$-functor $F : \mathcal{X} \to \mathcal{Y}$ *from one* $\mathcal{V}$-*category* $\mathcal{X}$ *to another* $\mathcal{Y}$ *is a function* $F : Ob(\mathcal{X}) \to Ob(\mathcal{Y})$, *together with an* $Ob(\mathcal{X})^2$-*indexed family of arrows in* $\mathcal{V}$,

---

[1]Unfortunately, the more obvious name of "topological category" has already been claimed for a class of generalizations of **Esp**.

[2]The reader will observe that we index the filtration in the opposite order from that commonly used in intersection cohomology.

$$F_{A,B} : \mathcal{X}(A, B) \to \mathcal{Y}(F(A), F(B)),$$

satisfying

$$
\begin{array}{ccc}
\mathcal{X}(A,B) \otimes \mathcal{X}(B,C) & \xrightarrow{\;\;\circ\;\;} & \mathcal{X}(A,C) \\[2mm]
\Big\downarrow{\scriptstyle F_{A,B} \otimes F_{B,C}} & & \Big\downarrow{\scriptstyle F_{A,C}} \\[2mm]
\mathcal{Y}(FA, FB) \otimes \mathcal{Y}(FB, FC) & \xrightarrow[\;\;\circ\;\;]{} & \mathcal{Y}(FA, FC)
\end{array}
$$

and

$$
\begin{array}{ccc}
 & \xrightarrow{\;Id_A\;} & \mathcal{X}(A,A) \\
I & & \Big\downarrow{\scriptstyle F_{A,A}} \\
 & \xrightarrow[\;Id_{FA}\;]{} & \mathcal{Y}(FA, FA)
\end{array}
$$

Similarly, one needs enriched natural transformations:

Definition 10.8 *A $\mathcal{V}$-natural transformation $\phi : F \Rightarrow G$ from one $\mathcal{V}$-functor $F : \mathcal{X} \to \mathcal{Y}$ to another $G : \mathcal{Y} \to \mathcal{Y}$ is an $Ob(\mathcal{X})$-indexed family of arrows in $\mathcal{V}$, $\phi_X : I \to \mathcal{Y}(F(X), G(X))$ satisfying*

$$
\begin{array}{ccc}
& \phi_A \otimes G_{A,B} & \\
I \otimes \mathcal{X}(A,B) \xrightarrow{\hspace{2cm}} & \mathcal{Y}(FA,GA) \otimes \mathcal{Y}(GA,GB)
\end{array}
$$

Diagram:

$I \otimes \mathcal{X}(A,B) \xrightarrow{\ \phi_A \otimes G_{A,B}\ } \mathcal{Y}(FA,GA) \otimes \mathcal{Y}(GA,GB)$

$\mathcal{X}(A,B)$, λ^{-1} , $\circ$, $\mathcal{Y}(FA,GB)$

ρ^{-1}

$\mathcal{X}(A,B) \otimes I \xrightarrow{\ F_{A,B} \otimes \phi_B\ } \mathcal{Y}(FA,FB) \otimes \mathcal{Y}(FB,GB)$, $\circ$

For the special cases we will use the obvious names: R-linear functors, R-linear natural transformations, complete R-linear functors, complete R-linear natural transformations, topologized functors, topologized natural transformations, stratified functors and stratified natural transformations.

In each case, given an enriched category, functor, or natural transformation, there is a corresponding "underlying" category obtained by taking as arrows maps from I to the hom-objects:

Definition 10.9 *Given a $\mathcal{V}$-category $\mathcal{X}$, the underlying category* **X** *has as objects the objects of $\mathcal{X}$, and as hom-sets the sets*

$$\mathcal{V}(I, \mathcal{X}(A,B))$$

with identity maps given by the map naming the enriched identity, and composition given by

$$fg \in \mathcal{V}(I, \mathcal{X}(A,C)) = \lambda_I(f \otimes g) \circ_{A}, B, C$$

for $f \in \mathcal{V}(I, \mathcal{X}(A,B))$ and $g \in \mathcal{V}(I, \mathcal{X}(B,C))$.

Given a $\mathcal{V}$-functor $F : \mathcal{X} \to \mathcal{Y}$ the underlying functor $F : \mathbf{X} \to \mathbf{Y}$ is given on objects by the same map, and on arrows by $F(f) = fF_{A,B}$ for $f \in \mathcal{V}(I, \mathcal{X}(A,B))$.

The components of the underlying natural transformation *of a V-natural transformation are simply the components as given, but now regarded as maps in the underlying category of the target of the underlying functors.*

For any symmetric monoidal category V, we can define an "underlying" (ordinary) category for any V category as follows: let $\mathbf{1}$ denote the V category with a single object, $*$, and $\mathbf{1}(*, *) = I$, the monoidal identity object in V. The underlying category of a V-category, $\mathcal{X}$, is then the category of V-functors and V-natural transformations from $\mathbf{1}$ to $\mathcal{X}$. In the cases of interest to us, this amounts to forgetting the underlying structure of the "hom-objects" to leave only the underlying "hom-set". Similarly, one may construct underlying (ordinary) functors and underlying (ordinary) natural transformations.

For this reason, it is easy to see that in all of the cases above, the cartesian product of two such underlying categories $\mathcal{X}$ and $\mathcal{Y}$ can be "enriched" by putting the natural structure of a V object on the cartesian product of the hom-sets to yield a V-category $\mathcal{X} \times \mathcal{Y}$.

In the case of topological and stratified categories this is a useful thing to do. However, for (complete) R-linear categories, this give the "wrong" result: the monoidal product on R-**mod** (or $\mathcal{X}_R$) is not the cartesian product. In particular, the monoidal structure on $R - \mathbf{mod}$ itself cannot be lifted to an R-linear functor from $R - \mathbf{mod} \times R - \mathbf{mod}$ to $R - \mathbf{mod}$.

To consider monoidal structures in the context of (complete) R-linear categories, it will be necessary to introduce a different product:

Definition 10.10 *The* Deligne product *of two R-linear categories (resp. complete R-linear categories) $\mathcal{X}$ and $\mathcal{Y}$, denoted $\mathcal{X} \boxtimes \mathcal{Y}$ (resp. $\mathcal{X} \hat{\boxtimes} \mathcal{Y}$), is the R-linear category with objects $Ob(\mathcal{X}) \times Ob(\mathcal{Y})$, and hom-objects given by*

$$\mathcal{X} \boxtimes \mathcal{Y}(<A, B>, <C, D>) = \mathcal{X}(A, C) \otimes \mathcal{Y}(B, D)$$

(resp.

$$\mathcal{X}\hat{\boxtimes}\mathcal{Y}(<A,B>,<C,D>) = \mathcal{X}(A,C)\hat{\otimes}\mathcal{Y}(B,D) \ ),$$

composition given by

$$[Id \otimes tw \otimes Id][\circ \otimes \circ],$$

and identities given by

$$Id_{<A,B>} = Id_A \otimes Id_B,$$

where $\hat{\otimes}$ denotes the $\mathfrak{m}$-adic completion of the tensor product over R in the case of complete R-linear categories.

Part II

Deformations

Chapter 11

Introduction

In Part II we discuss the deformation theory of monoidal categories brought to light by the author in collaboration with Crane [13], and the closely related infinitessimal deformation theories for monoidal functors and braided monoidal categories. We do not deal with the global aspects of categorical deformation theory whose invesigation was begun independently by Davydov [14] and in the special case of categorifications of group algebras and their quantum doubles by Crane and the author [13].

While an algebraist interested in categorical deformation theory in its own right may object to this omission, it is in keeping with the focus of this work on the categorical structures most intimately related to classical knot theory. As will become clear to the reader, the infinitesimal deformation theory of braided monoidal categories is intimately connected to the theory of Vassiliev invariants, which we also discuss in Part II.

The original motivation for the study of categorical deformations, however, was not to provide a functorial basis for Vassiliev theory. Rather, the author, with Crane, was motivated by the search for interesting examples of additive "Hopf categories" in the sense of Crane and Frenkel [12]. This original program remains incomplete, though recent advances in our understanding of categorical deformations, particularly with the discovery of deformation complexes for monoidal functors, and long-exact sequences relating various deformation complexes holds out hope for its completion. It is also possible that additive categories are the wrong setting for the theory of Hopf categories, and that the use of triangulated and derived categories is necessary, with some axioms holding only up to quasi-isomorphism (cf. recent work of Lyubashenko [38]).

Chapter 12

Definitions

We are concerned herein with the case of categories linear over some commutative ring R. As discussed in the previous chapter, the cartesian products of categories occurring in the definitions of Chapter 3 should be replaced with the product of Definition 10.10.

Now, given an R-linear category $\mathcal{C}$, and an R-algebra A, we can form a category $\mathcal{C} \otimes A$ by "extension of scalars":

$$Ob(\mathcal{C} \otimes A) = Ob(\mathcal{C})$$

and

$$Hom_{\mathcal{C} \otimes A}(X, Y) = Hom_{\mathcal{C}}(X, Y) \otimes_R A$$

with composition and $\otimes$ on maps extended by bilinearity. If A is an m-adically complete local ring, we can similarly construct $\mathcal{C} \widehat{\otimes} A$, by m-adically completing the Hom-sets and extending composition and $\otimes$ on maps by continuity.

For $A = R[\epsilon]/ < \epsilon^{n+1} >$ we denote $\mathcal{C} \otimes A$ by $\mathcal{C}^{(n)}$. For $A = R[[x]]$ we denote $\mathcal{C} \widehat{\otimes} A$ by $\mathcal{C}^{(\infty)}$.

Definition 12.1 *Given an R-linear semigroupal (resp. monoidal, braided monoidal, ribbon) category $\mathcal{C}$, an nth order deformation of the struc-*

ture on the category is a structure of the same type on $C^{(n)}$ whose structural functors are the extensions of those for C by bilinearity, and whose structural natural transformations reduce modulo ϵ to those of C.

Definition 12.2 *Given an R-linear semigroupal (resp. monoidal, braided monoidal, ribbon) category C a formal deformation of the structure on the category is a structure of the same type on $C^{(\infty)}$ whose structural functors are the extensions of those for C by bilinearity and continuity, and whose structural natural transformations reduce modulo $\mathfrak{m}$ to those of C.*

Definition 12.3 *The trivial deformation in either sense above is the deformation of C whose structural natural transformations are the images of those in C under extension of scalars. It is denoted $C^{(n)}_{\text{triv}}$ or $C^{(\infty)}_{\text{triv}}$ as appropriate.*

Definition 12.4 *Two deformations of monoidal categories in either sense above are equivalent if there are structural natural transformations that provide the identity functor with the structure of a semigroupal (resp. monoidal, braided monoidal, ribbon) functor between the deformations and which reduce to identity maps modulo ϵ or $\mathfrak{m}$.*

Definition 12.5 *Given an R-linear lax (resp. oplax, strong) semigroupal functor $F : C \to D$, a purely functorial n^{th} order deformation of the structure on the functor is a structure of the same type on $F^{(n)} : C^{(n)}_{\text{triv}} \to D^{(n)}_{\text{triv}}$ whose structural functors are the extensions of those for C by bilinearity, and whose structural natural transformations reduce modulo ϵ to those of F, where the source and target have the trivial deformation of semigroupal structure.*

Definition 12.6 *Given an R-linear lax (resp. oplax, strong) semigroupal functor $F : C \to D$, a purely functorial formal deformation of the structure on the functor is a structure of the same type on $F^{(\infty)} : C^{(\infty)}_{\text{triv}} \to D^{(\infty)}_{\text{triv}}$ whose structural functors are the extensions of those for*

C by bilinearity and continuity, and whose structural natural transformations reduce modulo ϵ to those of F, where the source and target have the trivial deformation of semigroupal structure.

Definition 12.7 *Two purely functorial deformations are* equivalent *if there is a semigroupal natural isomorphism between them which reduces to the identity natural isomorphism modulo ϵ or $\mathfrak{m}$.*

Definition 12.8 Fibred deformations *are defined similarly, but without the condition of triviality for the deformation of the source.* Total deformations *are defined in the same way, but without the condition of triviality on source or target.*

Definition 12.9 *Two fibred (resp. total) deformations are* equivalent *if there are structural natural transformations which provide the identity functor (resp. identity functors) of the underlying category (resp. categories) of the source (resp. source and target) of the deformation with the structure of a strong semigroupal functor between the deformationed semigroupal category structues, and which, moreover, reduce to identity maps modulo ϵ or $\mathfrak{m}$, provided the triangle (resp. square) of semigroupal functors formed by the deformed functors and the identity functor (resp. identity functors) equipped with these structure maps commutes up to a semigroupal natural isomorphism which, in turn, reduces modulo ϵ or $\mathfrak{m}$ to the identity natural isomorphism.*

Ribbon categories are the principal objects of interest for applications to low-dimensional topology, thanks to Shum's Coherence Theorem (Theorem 9.6) and the central role of framed links in 3- and 4-manifold topology. We will not, however, consider deformation theories for monoidal categories with duals. We do not consider such theories, because as a practical matter we usually begin by deforming a rigid symmetric tensor category, and we have:

Theorem 12.10 *Any n^{th} order or formal braided monoidal deformation of a rigid symmetric K-linear tensor category C for K any field (char $k \neq 2$) admits a unique tortile structure, and is thus a tortile deformation.*

This result follows from essentially the same proof as the theorem of Deligne on braided monoidal deformations of Tannakian categories (cf. [61]).

Most of the deformation theory for braided monoidal categories can be constructed from the deformation theory for monoidal categories and that for strong monoidal functors by use of the following results of Joyal and Street [30]:

Definition 12.11 *A multiplication on a monoidal category C is a strong monoidal functor $(\Phi : C \times C \to C, \tilde{\Phi}, \Phi_0)$ (usually denoted Φ by abuse of notation), together with monoidal natural isomorphisms $\mathfrak{r} : \Phi(Id_C, I) \Rightarrow Id_C$ and $\mathfrak{l} : \Phi(I, Id_C) \Rightarrow Id_C$.*

Theorem 12.12 *In a monoidal category C, a family of arrows*

$$\sigma_{A,B} : A \otimes B \longrightarrow B \otimes A$$

is a braiding if and only if the following define a multiplication Φ on C: $\Phi = \otimes$, $\Phi_0 = \rho_I^{-1}$, $\mathfrak{r} = \rho$, $\mathfrak{l} = \lambda$ and

$$\tilde{\Phi}_{(A,A'),(B,B')} =$$
$$\lceil (1 \otimes \sigma) \otimes 1 \rceil : (A \otimes A') \otimes (B \otimes B') \longrightarrow (A \otimes B) \otimes (A' \otimes B').$$

Conversely, we have

Theorem 12.13 *For any multiplication Φ on a monoidal category C, a braiding σ for C is defined by the commutative diagram of Figure 12.1. The multiplication obtained from this braiding via Theorem 12.12 is isomorphic (in the obvious sense) to Φ. If τ is any braiding on C and the multiplication Φ is obtained from τ by the construction of Theorem 12.12, then $\sigma = \tau$.*

As is observed in [30], this last result is an analogue of the old result of Eckmann and Hilton [20], usually remembered as "A group in **Groups** is an abelian group," although it actually applies to monoids.

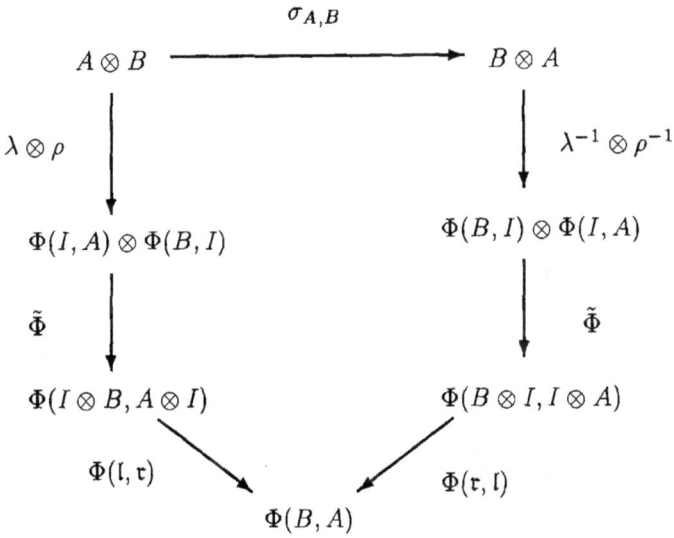

Figure 12.1: The Braiding Associated to a Multiplication

Chapter 13

Deformation Complexes of Semigroupal Categories and Functors

We can now conveniently define a cochain complex associated to any semigroupal category or semigroupal functor:

Definition 13.1 *The* deformation complex *of a semigroupal category* $\mathcal{C}, \otimes, \alpha$ *is the cochain complex* $X^{\bullet}(\mathcal{C}), \delta$ *where*

$$X^n(\mathcal{C}) = \text{Nat}(^n\otimes, \otimes^n)$$

and

$$\delta(\phi)_{A_0,\ldots,A_n} = \lceil A_0 \otimes \phi_{A_1,\ldots,A_n} \rceil + \sum_{i=1}^{n}(-1)^i \lceil \phi_{A_0,\ldots,A_{i-1}\otimes A_i,\ldots,A_n} \rceil$$
$$+ (-1)^{n+1}\lceil \phi_{A_0,\ldots,A_{n-1}} \otimes A_n \rceil.$$

Definition 13.2 *The* deformation complex *of a lax semigroupal functor* $(F : \mathcal{C} \to \mathcal{D}, \tilde{F})$ *is the cochain complex*

149

$$(X^\bullet(F), \delta),$$

where

$$X^n(F) = \mathrm{Nat}({}^n\otimes(F^n), F(\otimes^n))$$

and

$$
\begin{aligned}
\delta(\phi)_{A_0,\dots,A_n} = & \ \lceil F(A_0) \otimes \phi_{A_1,\dots,A_n} \rceil \\
& + \sum_{i=1}^{n}(-1)^i \lceil \phi_{A_0,\dots,A_{i-1}\otimes A_i,\dots,A_n} \rceil \\
& + (-1)^{n+1} \lceil \phi_{A_0,\dots,A_{n-1}} \otimes F(A_n) \rceil.
\end{aligned}
$$

Definition 13.3 *The* deformation complex *of an oplax or strong semi-groupal functor* $(F : \mathcal{C} \to \mathcal{D}, \tilde{F})$ *is the cochain complex*

$$(X^\bullet(F), \delta)$$

where

$$X^n(F) = \mathrm{Nat}(F({}^n\otimes), \otimes^n(F^n))$$

and

$$
\begin{aligned}
\delta(\phi)_{A_0,\dots,A_n} = & \ \lceil F(A_0) \otimes \phi_{A_1,\dots,A_n} \rceil \\
& + \sum_{i=1}^{n}(-1)^i \lceil \phi_{A_0,\dots,A_{i-1}\otimes A_i,\dots,A_n} \rceil \\
& + (-1)^{n+1} \lceil \phi_{A_0,\dots,A_{n-1}} \otimes F(A_n) \rceil.
\end{aligned}
$$

The motivation for these definitions can be found in [13] and [63], or can be readily discovered by the reader by computing by hand the conditions on the term $\alpha^{(1)}$ in a first order deformation $\alpha^{(0)} + \alpha^{(1)}\epsilon$

of a semigroupal category $(\mathcal{C}, \otimes, \alpha = \alpha^{(\prime)})$, and on the term $\tilde{F}^{(1)}$ in a purely functorial deformation $\tilde{F}^{(0)} + \tilde{F}^{(1)}\epsilon$ of a semigroupal functor $F : \mathcal{C} \to \mathcal{D}, \tilde{F} = \tilde{F}^{(0)}$.

In [13] it is shown that

Theorem 13.4 *The first-order deformations of a semigroupal category $\mathcal{C}$ are classified up to equivalence by $H^3(\mathcal{C})$.*

sketch of proof: Consider two first-order deformations $\tilde{\alpha} = \alpha + \alpha^{(1)}\epsilon$ and $\hat{\alpha} = \alpha + a^{(1)}\epsilon$ of $\mathcal{C}$. Consider also a semigroupal functor whose underlying functor is the identity functor and whose structural transformation is of the form

$$1_{A \otimes B} + \phi_{A,B}\epsilon : A \otimes B \to A \otimes B.$$

Now, write out the coherence condition for semigroupal functors in this case, and look at the degree 1 terms. The resulting equation is nothing more than

$$\alpha^{(1)} - a^{(1)} = \delta(\phi).$$

□

In [63] it is shown that

Theorem 13.5 *The purely functorial first-order deformations of a semigroupal functor $F : \mathcal{C} \to \mathcal{D}$ are classified up to equivalence by $H^2(F)$.*

The proof is similar to that of the previous theorem, and may be readily reconstructed by the reader, or found in [63].

Obstructions to extending n^{th} order deformation to $(n + 1)^{st}$ order deformations are discussed in either case in [13] and [63], and we will return to this matter later.

First, however, we wish to consider how to use the deformation complexes already defined to deal with the cases of fibred and total deformations and deformations of braided monoidal categories.

Although the degree and type of functoriality properties satisfied by the deformation complexes for semigroupal categories and functors is an open question, two rather comforting results hold:

Theorem 13.6 *If C and D are two semigroupally equivalent categories, then the deformation complexes $X^{\bullet}(C)$ and $X^{\bullet}(C)$ are isomorphic.*

and

Theorem 13.7 *If $F, G : C \to D$ are semigroupal functors, any semigroupal natural isomorphism induces an isomorphism between the deformation complexes $X^{\bullet}(F)$ and $X^{\bullet}(G)$.*

The proofs are rather obvious and are left to the reader. The only ticklish bits, once one writes down the obvious maps on the cochain groups, are showing that they collectively form a cochain map and showing that the maps induced in each direction by the equivalence of categories are actually inverse to each other. In both cases the semigroupal property of the natural isomorphism or of the natural isomorphisms defining the equivalence is crucial.

Chapter 14

Some Useful Cochain Maps

The compositions of natural transformations with functors allow us to induce two cochain maps whenever we have a semigroupal functor $F : \mathcal{C} \to \mathcal{D}$ (whether lax, oplax, or strong):

$$\lceil F(-) \rceil : X^{\bullet}(\mathcal{C}) \to X^{\bullet}(F)$$

and

$$\lceil (-)_{F\bullet} \rceil : X^{\bullet}(\mathcal{D}) \to X^{\bullet}(F).$$

We should note that $X^{\bullet}(\mathcal{C}) = X^{\bullet}(Id_{\mathcal{C}})$. The cochain maps just defined are, in fact, special cases of more general families of cochain maps defined for any composable pair of functors

$$\mathcal{C} \xrightarrow{F} \mathcal{D} \xrightarrow{G} \mathcal{E}$$

namely,

$$\lceil G(-) \rceil : X^{\bullet}(F) \to X^{\bullet}(G(F))$$

and

$$\lceil (-)_{F \bullet} \rceil : X^\bullet(G) \to X^\bullet(G(F)).$$

We will not here pursue the use of these more general notions, save to note in passing that they provide long-exact sequences relating the cohomology of a composite functor with the cohomology of its factors and some sort of "relative cohomology" which, in the appropriate dimensions, will measure the extent to which deformations of the composite are not accounted for by the deformations of the factors.

To consider deformations of braided monoidal categories the following is also useful: given any K-linear semigroupal category $\mathcal{C}$, there is a "diagonal" cochain map

$$\Delta : X^\bullet(\mathcal{C}) \to X^\bullet(\mathcal{C} \boxtimes \mathcal{C})$$

given by:

$$\Delta(\phi) = \phi \boxtimes \lceil Id \rceil + \lceil Id \rceil \boxtimes \phi \ .$$

These cochain maps allow us to assemble the simpler deformation complexes for semigroupal categories and semigroupal functors into complexes whose cohomology is related to more general types of deformations.

Recall the construction of a cone over a cochain map:

Definition 14.1 *Given a map of cochain complexes* $u^\bullet : A^\bullet \to B^\bullet$, *the cone on* $u^\bullet$ *is the cochain complex*

$$(C_u^\bullet, d_u) = \left(B^\bullet \oplus A^{\bullet+1}, \begin{bmatrix} d_B & 0 \\ u & -d_A \end{bmatrix} \right).$$

Here we adopt the convention that elements of direct sums are written as row vectors with entries in the summands, and that arrays of maps act on the right by matrix multiplication (with the action of maps in lieu of scalar multiplication). Note that this is consistent with our notational convention: maps *act* on the right on elements (improperly) thought of as maps, unless parentheses denoting application intervene.

Chapter 15

First Order Deformations

Let us now consider the problem of classifying first order fibred deformations of semigroupal functors. If we have a lax (resp. oplax, strong) semigroupal functor $[F, \tilde{F}] : (\mathcal{C}, \otimes, \alpha) \to (\mathcal{D}, \otimes, a)$, and we replace $\tilde{F}^{(0)} = \tilde{F}$ with $\tilde{F}^{(0)} + \tilde{F}^{(1)}\epsilon$ and $\alpha^{(0)} = \alpha$ with $\alpha^{(0)} + \alpha^{(1)}\epsilon$ for $\epsilon^2 = 0$, the conditions for the new coherence diagrams to commute become

$$\delta(\alpha^{(1)}) = 0$$

and

$$\delta(\tilde{F}^{(1)}) + \lceil F(\alpha) \rceil = 0,$$

as can be verified readily by computing the ϵ-degree 1 terms going around the pentagon and hexagon coherence diagrams.

It then follows directly that the pair $[\tilde{F}^{(1)}, \alpha^{(1)}]$ is a 2-cocycle in

$$(C^{\bullet}_{\lceil F(-) \rceil}, d_{-\lceil F(-) \rceil}).$$

Now, consider the condition that two such 2-cocycles $[\tilde{F}_1^{(1)}, \alpha_1^{(1)}]$ and $[\tilde{F}_2^{(1)}, \alpha_2^{(1)}]$ are equivalent. Let $F_1 : \mathcal{C}_1 \to \mathcal{D}$ and $F_2 : \mathcal{C}_2 \to \mathcal{D}$ denote the semigroupal functors from the corresponding deformations (suppressing here the naming of structural maps). In particular, there is a structure map which makes $Id_{\mathcal{C} \otimes R[\epsilon]/<\epsilon^2>}$ into a (necessarily strong) semigroupal functor and which reduces modulo ϵ to the identity natural transformation. Second, there is a semigroupal natural isomorphism ψ from F_1 to $F_2(\mathfrak{I})$ which reduces modulo ϵ to the identity, where $\mathfrak{I}$ is the identity functor on $\mathcal{C} \otimes R[\epsilon]/ < \epsilon^2 >$ made into a semigroupal functor by given structure map.

Denoting the structural map for $Id_{\mathcal{C} \otimes R[\epsilon]/<\epsilon^2>}$ by $id + \iota^{(1)}\epsilon$ and letting $\psi = id + \psi^{(1)}\epsilon$, the coherence conditions become

$$[\tilde{F} + \tilde{F}^{(1)}_{2A,B}\epsilon]([id + \iota^{(1)}_{A,B}\epsilon](id_{F(A\otimes B)} + \psi^{(1)}_{A\otimes B}\epsilon)) =$$
$$[[id_{F(A)} + \psi^{(1)}_A\epsilon] \otimes [id_{F(B)} + \psi^{(1)}_B\epsilon]](\tilde{F} + \tilde{F}^{(1)}_{2A,B}\epsilon)$$

and

$$[id_{F(A)} \otimes [id_{F(B\otimes C)} + \iota^{(1)}_{B,C}\epsilon]]$$
$$([id_{F(A\otimes[B\otimes C])} + \iota^{(1)}_{A,B\otimes C}\epsilon](F(\alpha + \alpha^{(1)}_{1A,B,C}\epsilon))) =$$
$$[F(\alpha + \alpha^{(1)}_{2A,B,C}\epsilon)]([[id_{F(A\otimes B)} + \iota^{(1)}_{A,B}] \otimes id_{F(C)}]$$
$$([id_{F([A\otimes B]\otimes C)} + \iota^{(1)}_{A\otimes B,C}\epsilon])) \ .$$

Using the bilinearity of composition and $\otimes$, the coherence conditions on the original maps, and the condition $\epsilon^2 = 0$, these readily reduce to

$$\tilde{F}^{(1)}_1 - \tilde{F}^{(1)}_2 = \iota^{(1)} - \delta(\psi^{(1)})$$

and

$$\alpha^{(1)}_1 - \alpha^{(1)}_2 = \delta(\iota^{(1)}) \ .$$

We have thus demonstrated

Theorem 15.1 *The first order fibred deformations of a semigroupal functor $F : \mathcal{C} \to \mathcal{D}$ are classified up to equivalence by the third cohomology of the cone $C^{\bullet}_{\lceil F(-) \rceil} = X^{\bullet}_{\mathrm{fibred}}(F)$.*

A similar analysis replacing $\lceil F \rceil : X^\bullet(\mathcal{C}) \to X^\bullet(F)$ with

$$\lceil F(p_1) \rceil - \lceil (p_2)_{F^\bullet} \rceil : X^\bullet(\mathcal{C}) \oplus X^\bullet(\mathcal{D}) \to X^\bullet(F)$$

shows that

Theorem 15.2 *The first order total deformations of a semigroupal functor* $F : \mathcal{C} \to \mathcal{D}$ *are classified up to equivalence by the third cohomology of the cone* $C^\bullet_{\lceil F(p_1) \rceil - \lceil (p_2)_{F^\bullet} \rceil} = X^\bullet_{\mathrm{total}}(F)$.

The case of total deformations of a multiplication (or equivalently, deformations of a braided monoidal category) presents another subtlety: the source and target must be deformed in tandem.

Proposition 15.3 *If* $\mathcal{C}^{(n)}, \otimes, \alpha^{(0)} + \alpha^{(1)} \epsilon + \ldots + \alpha^{(n)} \epsilon^n$ *is an* n^{th}*-order deformation of* $\mathcal{C}, \otimes, \alpha$ *and*

$$\beta^{(k)} = \sum_{i=0}^{k} \alpha^{(i)} \boxtimes \alpha^{(k-i)} \ ,$$

then $[\mathcal{C} \boxtimes \mathcal{C}]^{(n)}, \otimes \boxtimes \otimes, \beta^{(0)} + \beta^{(1)} \epsilon + \ldots \beta^{(n)} \epsilon^n$ *is an* n^{th}*-order deformation of* $\mathcal{C} \boxtimes \mathcal{C}, \otimes \boxtimes \otimes, \alpha \boxtimes \alpha$. *We call this deformation the* diagonal deformation *of* $\mathcal{C} \boxtimes \mathcal{C}$.

proof: Observe first that $\mathcal{C} \boxtimes \mathcal{C}$ is defined with respect to the commutative ring R, and that $[\mathcal{C} \boxtimes_R \mathcal{C}]^{(n)}$ is canonically isomorphic to $\mathcal{C}^{(n)} \boxtimes_{R[\epsilon]/<\epsilon^n>} \mathcal{C}^{(n)}$.

The diagonal deformation is then simply the $R[\epsilon]/<\epsilon^n>$-linearized version of the diagonal semigroupal structure induced on $\mathcal{C}^{(n)} \times \mathcal{C}^{(n)}$ by the (deformed) semigroupal structure on $\mathcal{C}^{(n)}$. The formula for the $\beta^{(k)}$'s is derived by simply collecting terms according to their degree in ϵ. $\square$

A similar result holds for formal deformations.

Definition 15.4 *A* coarse deformation *of a multiplication is a total deformation of the semigroupal functor such that the deformation of the source* $C \boxtimes C$ *is the diagonal deformation induced by the deformation of the target.   A deformation* of a multiplication *(and thus of a braided monoidal category) is a coarse deformation which is equipped with natural isomorphisms as required to make it into a multiplication.*

We will consider the behavior of units in general in Chapter 17, so we here confine ourselves to consider the appropriate deformation complex for coarse deformations of multiplications:

Consider the composite cochain map

$$\phi : X^\bullet(C) \xrightarrow{(\Delta, Id)} X^\bullet(C \boxtimes C) \oplus X^\bullet(C) \xrightarrow{\lceil \Phi(p_1) \rceil - \lceil (p_2)_{\Phi} \bullet \rceil} X^\bullet(\Phi).$$

An argument similar to that given above for fibred deformations shows that:

Theorem 15.5 *The first order coarse deformations of a multiplication* $\Phi : C \boxtimes C \to C$ *are classified up to equivalence by the third cohomology of the cone* $C_\phi^\bullet = X_{\mathrm{coarse}}^\bullet(\Phi).$

Since ϕ is defined as a composite, something more remains to be said: if we consider our cochain complexes as objects in the homotopy category $K^+(R)$ or the derived category $D^+(R)$, the octahedral property ensures the existence of an exact triangle relating $X_{\mathrm{coarse}}^\bullet(\Phi)$, $X_{\mathrm{total}}^\bullet(\Phi)$ and $C_{(\Delta, Id)}$, and thus of a long-exact sequence in cohomology.

Chapter 16

Obstructions and the Cup Product and Pre-Lie Structures on $X^\bullet(F)$

The cochain complex associated to any of the types of semigroupal functors shares many of the properties of the Hochschild complex of an associative algebra A with coefficients in A, which were described by Gerstenhaber [23, 24, 25]. Indeed, in Chapter 21 we will see that the Hochschild complex, with all of the structure discovered by Gerstenhaber, is a special case of our construction.

In particular, we have two products defined on cochains. The first, the cup product,

$$ - \cup - : X^n(F) \times X^m(F) \to X^{n+m}(F) \,, $$

is given by

$$ G \cup H_{A_1,\dots A_{n+m}} = \lceil G_{A_1,\dots,A_n} \otimes H_{A_{n+1},\dots,A_{n+m}} \rceil \,. $$

The second, the composition product,

$$ \langle -, - \rangle : X^n(F) \times X^m(F) \to X^{n+m-1}(F) $$

is given by

$$\langle G, H \rangle_{A_1, \dots A_{n+m-1}} =$$
$$\sum (-1)^{mi} \lceil F(A_1) \otimes \dots F(A_i) \otimes H_{A_{i+1}, \dots, A_{i+n}} \otimes F(A_{i+n+1}) \otimes$$
$$\dots \otimes F(A_{n+m-1})(G_{A_1, \dots, A_i, A_{i+1} \otimes \dots \otimes A_{i+n}, A_{i+n+1}, \dots A_{n+m-1}}) \rceil$$

in the case of oplax and strong semigroupal functors, and by

$$\langle G, H \rangle_{A_1, \dots A_{n+m-1}} =$$
$$\sum (-1)^{mi} \lceil (G_{A_1, \dots, A_i, A_{i+1} \otimes \dots \otimes A_{i+n}, A_{i+n+1}, \dots A_{n+m-1}})$$
$$F(A_1) \otimes \dots F(A_i) \otimes H_{A_{i+1}, \dots, A_{i+n}} \otimes F(A_{i+n+1}) \otimes$$
$$\dots \otimes F(A_{n+m-1}) \rceil$$

in the case of lax semigroupal functors.

Proposition 16.1 *The product* $\langle -, - \rangle$ *comes from a "pre-Lie system", in the terminology of Gerstenhaber [23], given by*

$$\langle G, H \rangle^{(i)}_{A_1, \dots A_{n+m-1}} =$$
$$\lceil F(A_1) \otimes \dots F(A_i) \otimes H_{A_{i+1}, \dots, A_{i+n}} \otimes F(A_{i+n+1}) \otimes$$
$$\dots \otimes F(A_{n+m-1})(G_{A_1, \dots, A_i, A_{i+1} \otimes \dots \otimes A_{i+n}, A_{i+n+1}, \dots A_{n+m-1}}) \rceil$$

in the case of oplax and strong semigroupal functors, and by

$$\langle G, H \rangle^{(i)}_{A_1, \dots A_{n+m-1}} =$$
$$\lceil (G_{A_1, \dots, A_i, A_{i+1} \otimes \dots \otimes A_{i+n}, A_{i+n+1}, \dots A_{n+m-1}}) F(A_1) \otimes \dots$$
$$F(A_i) \otimes H_{A_{i+1}, \dots, A_{i+n}} \otimes F(A_{i+n+1}) \otimes \dots \otimes F(A_{n+m-1}) \rceil$$

in the case of lax semigroupal functors, where in either case $X^n(F)$ *has degree* $n - 1$.

proof: First, note that the ambiguities of parenthesization in the semi-groupal products in this definition are rendered irrelevant by the $\lceil \ \rceil$ on each term, by virtue of the coherence theorems for semigroupal functors.

It is obvious that the product is given by a sum of these terms with the correct signs for the construction of a Lie bracket from a pre-Lie system, so actually the content of the proposition is that the $\langle -, - \rangle^{(i)}$'s satisfy the definition of a pre-Lie system. That is, for $G \in X^m(F)$, $H \in X^n(F)$ and $K \in X^p(F)$, we have

$$\langle\langle G, H\rangle^{(i)}, K\rangle^{(j)} = \begin{cases} \langle\langle G, K\rangle^{(j)}, H\rangle^{(i+p-1)} & \text{if } 0 \leq j \leq i-1 \\ \langle G, \langle H, K\rangle^{(j-i)}\rangle^{(i)} & \text{if } i \leq j \leq n \end{cases}$$

(recall that a k-chain has degree $k - 1$).

This is a simple computational check. One must remember that naturality will allow one to commute the prolongations of K and H in verifying the first case. $\square$

Now, suppose we have an $M - 1^{st}$ order deformation

$$\tilde{\alpha} = \alpha^{(0)} + \alpha^{(1)}\epsilon + \ldots + \alpha^{(M-1)}\epsilon^{M-1} \ .$$

As was shown in [13], the obstruction to extending this to an Mth order deformation is the 4-cochain

$$\omega^{(M)}_{A,B,C,D} = \sum_{\substack{i+j=M \\ 0 \leq i,j < M}} \lceil \alpha^{(i)}_{A \otimes B, C, D} \alpha^{(j)}_{A,B,C \otimes D} \rceil$$

$$- \sum_{\substack{i+j+k=M \\ 0 \leq i,j,k < M}} \lceil [\alpha_{A,B,C} \otimes D]\alpha^{(j)}_{A, B \otimes C, D}[A \otimes \alpha^{(k)}_{B,C,D}]\rceil.$$

The deformation extends precisely when this cochain is a cobound-ary, in which case $\alpha^{(M)}$ may be any solution to $\delta(\alpha^{(M)}) = \omega^{(M)}$.

What was heretofore missing was

Theorem 16.2 *For all M, the obstruction $\omega^{(M)}$ is a 4-cocycle. Thus, an $(M - 1)^{st}$ order deformation extends to an Mth order deformation if and only if the cohomology class $[\omega^{(M)}] \in H^4(\mathcal{C})$ vanishes.*

proof: The proof is essentially computational, and it is thus desirable to have briefer notation for many of its key ingredients.

Given a 3-cochain $\phi_{A,B,C}$, we denote the summands of its coboundary by

$$
\begin{aligned}
\partial_0 \phi_{A,B,C,D} &= A \otimes \phi_{B,C,D} \\
\partial_1 \phi_{A,B,C,D} &= \phi_{A \otimes B, C, D} \\
\partial_2 \phi_{A,B,C,D} &= \phi_{A, B \otimes C, D} \\
\partial_3 \phi_{A,B,C,D} &= \phi_{A, B, C \otimes D} \\
\partial_4 \phi_{A,B,C,D} &= \phi_{A,B,C} \otimes D
\end{aligned}
$$

Similarly, for a 4-cochain $\psi_{A,B,C,D}$ we let

$$
\begin{aligned}
\underline{\partial}_0 \psi_{A,B,C,D,E} &= A \otimes \psi_{B,C,D,E} \\
\underline{\partial}_1 \psi_{A,B,C,D,E} &= \psi_{A \otimes B, C, D, E} \\
\underline{\partial}_2 \psi_{A,B,C,D,E} &= \psi_{A, B \otimes C, D, E} \\
\underline{\partial}_3 \psi_{A,B,C,D,E} &= \psi_{A, B, C \otimes, D, E} \\
\underline{\partial}_4 \psi_{A,B,C,D,E} &= \psi_{A, B, C, D \otimes E} \\
\underline{\partial}_5 \psi_{A,B,C,D,E} &= \psi_{A, B, C, D} \otimes E
\end{aligned}
$$

(We include the underline stroke only for ease of reading, not out of any logical necessity.)

We then have

$$
\delta(\phi)_{A,B,C,D} = \sum_{i=0}^{4} (-1)^{i+1} \partial_i \phi_{A,B,C,D}
$$

for 3-cochains ϕ, and

$$\delta(\psi)_{A,B,C,D,E} = \sum_{i=0}^{5}(-1)^{i+1}\underline{\partial}_i\psi_{A,B,C,D,E}$$

for 4-cochains ψ.

In this notation the obstruction cochain ω^M becomes

$$\omega^{(M)} = \sum_{\substack{i+j=M \\ 0 \le i,j < M}} \lceil \partial_1 \alpha^{(i)} \partial_3 \alpha^{(j)} \rceil - \sum_{\substack{i+j+k=M \\ 0 \le i,j,k < M}} \lceil \partial_4 \alpha^{(i)} \partial_2 \alpha^{(j)} \partial_0 \alpha^{(k)} \rceil \; ,$$

while the vanishing of the obstruction $\omega^{(N)}$ (for $N < M$) becomes

$$
\begin{aligned}
0 &= \delta\alpha^{(N)} + \omega^{(N)} \\
&= \sum_{\substack{i+j=N \\ 0 \le i,j \le N}} \lceil \partial_1 \alpha^{(i)} \partial_3 \alpha^{(j)} \rceil - \sum_{\substack{i+j+k=N \\ 0 \le i,j,k \le N}} \lceil \partial_4 \alpha^{(i)} \partial_2 \alpha^{(j)} \partial_0 \alpha^{(k)} \rceil \; .
\end{aligned}
$$

We wish to show that

$$\sum_{i=0}^{5}(-1)^{i+1}\underline{\partial}_i\omega^{(M)}_{A,B,C,D,E} = 0 \; .$$

Observe that $\omega^{(1)} = 0$ and $\delta(\alpha^{(1)}) = 0$, so we may proceed by induction under the assumption that $\omega^{(N)}$ and $\alpha^{(N)}$ satisfy

$$\delta(\alpha^{(N)}) + \omega^{(N)} = 0$$

for $N < M$.

It is convenient to picture the summands of the left-hand side in terms of compositions of maps along the boundaries of faces of the "associahedron" (or 3-dimensional Stasheff polytope) [52] given in Figure 16.1.

Suppose we have an $(M-1)^{st}$ order deformation of a semigroupal category with structure map. Observe that each summand

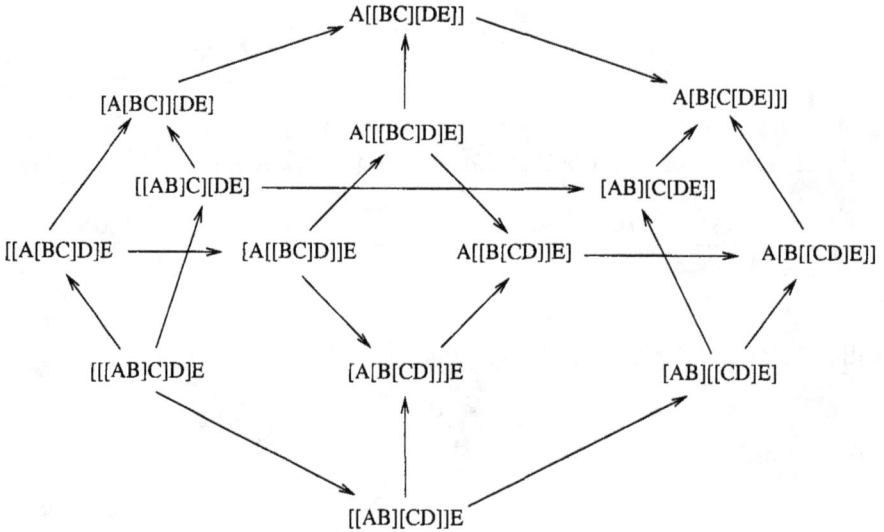

Figure 16.1: The Associahedron

$$\underline{\partial}_i \omega^{(M)}_{A,B,C,D,E}$$

essentially represents the sum of all composites with total degree M along the three-edge directed path minus the sum of all composites with total degree M along the two-edge directed path on the boundary of one of the pentagonal faces. (Here, degree refers to the power of ϵ whose coefficient is given by the composite.)

This is "essentially" the content of each summand, but one must remember that the context ⌈ ⌉ is not contentless—the summands are actually composites of the differences just described with various structure maps (prolongations of $\alpha^{(0)}$) with the property that all sources are $[[[A \otimes B] \otimes C] \otimes D] \otimes E$ and all targets are $A \otimes [B \otimes [C \otimes [D \otimes E]]]$.

The odd-index summands correspond to the pentagonal faces on the bottom of the associahedron as shown in Figure 16.1, while the even-index summands correspond to those on the top. The square faces of the associahedron correspond to families of naturality squares, one for each possible pair of degrees.

In fact, it will suffice to compute $[\underline{\partial}_1 + \underline{\partial}_3 + \underline{\partial}_5](\omega^{(M)})$:

Lemma 16.3 *Suppose for all* $N < M$ *we have* $\delta(\alpha^{(N)}) + \omega^{(N)} = 0$. *Then*

$$[\underline{\partial}_1 + \underline{\partial}_3 + \underline{\partial}_5](\omega^{(M)}) =$$
$$\sum_{\substack{i+j+k=M \\ 0 \le i,j,k < M}} \lceil \underline{\partial}_1 \partial_1 \alpha^{(i)} \underline{\partial}_1 \partial_3 \alpha^{(j)} \underline{\partial}_4 \partial_3 \alpha^{(k)} \rceil$$
$$- \sum_{\substack{i+j+k+l+m+n=M \\ 0 \le i,j,k,l,m,n < M}} \lceil \underline{\partial}_5 \partial_4 \alpha^{(i)} \underline{\partial}_5 \partial_2 \alpha^{(j)} \underline{\partial}_5 \partial_0 \alpha^{(k)} \underline{\partial}_3 \partial_2 \alpha^{(l)} \underline{\partial}_3 \partial_0 \alpha^{(m)} \underline{\partial}_0 \partial_0 \alpha^{(n)} \rceil .$$

Before proving this lemma, we should note why this lemma suffices to complete the proof of the theorem: the lemma and calculation by which it is derived are precisely dual to a corresponding statement and derivation concerning $[\underline{\partial}_0 + \underline{\partial}_2 + \underline{\partial}_4](\omega^{(M)})$. The value derived for this last expression is

$$\sum_{\substack{i+j+k=M \\ 0 \le i,j,k < M}} \lceil \underline{\partial}_1 \partial_1 \alpha^{(i)} \underline{\partial}_4 \partial_1 \alpha^{(j)} \underline{\partial}_4 \partial_3 \alpha^{(k)} \rceil$$

$$- \sum_{\substack{i+j+k+l+m+n=M \\ 0 \le i,j,k,l,m,n < M}} \lceil \underline{\partial}_5 \partial_4 \alpha^{(i)} \underline{\partial}_2 \partial_4 \alpha^{(j)} \underline{\partial}_2 \partial_2 \alpha^{(k)} \underline{\partial}_0 \partial_4 \alpha^{(l)} \underline{\partial}_0 \partial_2 \alpha^{(m)} \underline{\partial}_0 \partial_0 \alpha^{(n)} \rceil .$$

Once coincidences of different names for the same map (all of which may be read off from the associahedron) are taken into account, this expression differs from that computed in the lemma only in the third and fourth factors of the composites in the second summation. The terms, however, may be matched one-to-one by swapping the indices k and l into pairs that are equal by virtue of naturality, thus completing the proof.

Thus, it suffices to prove Lemma 16.3.

proof of Lemma 16.3

We begin by computing part of the sum:

$$[\underline{\partial}_5 + \underline{\partial}_3](\omega^{(M)}) = \sum_{\substack{i+j=M \\ 0 \le i,j < M}} \lceil \underline{\partial}_5\partial_1\alpha^{(i)} \underline{\partial}_5\partial_3\alpha^{(j)} \rceil$$

$$- \sum_{\substack{i+j+k=M \\ 0 \le i,j,k < M}} \lceil \underline{\partial}_5\partial_4\alpha^{(i)} \underline{\partial}_5\partial_2\alpha^{(j)} \underline{\partial}_5\partial_0\alpha^{(k)} \rceil$$

$$+ \sum_{\substack{i+j=M \\ 0 \le i,j < M}} \lceil \underline{\partial}_3\partial_1\alpha^{(i)} \underline{\partial}_3\partial_3\alpha^{(j)} \rceil$$

$$- \sum_{\substack{i+j+k=M \\ 0 \le i,j,k < M}} \lceil \underline{\partial}_3\partial_4\alpha^{(i)} \underline{\partial}_3\partial_2\alpha^{(j)} \underline{\partial}_3\partial_0\alpha^{(k)} \rceil.$$

Again, it is important to recall the meaning of the context $\lceil\ \rceil$: all of the summands here are maps from $[[[A \otimes B] \otimes C] \otimes D] \otimes E$ to $A \otimes [B \otimes [C \otimes [D \otimes E]]]$, involving implied structure maps from the undeformed category as factors (of the form $\partial_i\partial_j\alpha^{(0)}$ for various i and j).

Now, observe that $\underline{\partial}_5\partial_3 = \underline{\partial}_3\partial_4$, these two expressions being names for the same edge of the associahedron, but viewed from different faces. Since all of the indices occurring in either of these two expressions are less than M, we may use the hypothesized vanishing condition to replace each occurrence of either with summands corresponding to paths around the other face of the associahedron. In this way we obtain

$$[\underline{\partial}_5 + \underline{\partial}_3](\omega^{(M)}) =$$
$$- \sum_{\substack{i+j+k+l=M \\ 0 < i \\ 0 < k+l}} \lceil \underline{\partial}_5\partial_1\alpha^{(i)} \underline{\partial}_3\partial_4\alpha^{(j)} \underline{\partial}_3\partial_2\alpha^{(k)} \underline{\partial}_3\partial_0\alpha^{(l)} \rceil$$

$$+ \sum_{\substack{i+j+k=M \\ 0<i}} \lceil \underline{\partial}_5\partial_1\alpha^{(i)}\underline{\partial}_3\partial_1\alpha^{(j)}\underline{\partial}_3\partial_3\alpha^{(k)} \rceil$$

$$- \sum_{\substack{i+j+k=M \\ 0\le i,j,k<M}} \lceil \underline{\partial}_5\partial_4\alpha^{(i)}\underline{\partial}_5\partial_2\alpha^{(j)}\underline{\partial}_5\partial_0\alpha^{(k)} \rceil$$

$$+ \sum_{\substack{i+j=M \\ 0\le i,j<M}} \lceil \underline{\partial}_3\partial_1\alpha^{(i)}\underline{\partial}_3\partial_3\alpha^{(j)} \rceil$$

$$+ \sum_{\substack{i+j+k+l=M \\ 0<i \\ 0<k+l}} \lceil \underline{\partial}_5\partial_1\alpha^{(i)}\underline{\partial}_3\partial_4\alpha^{(j)}\underline{\partial}_3\partial_2\alpha^{(k)}\underline{\partial}_3\partial_0\alpha^{(l)} \rceil$$

$$- \sum_{\substack{i+j+k+l+m=M \\ i+j+k<M \\ 0\le l,m<M}} \lceil \underline{\partial}_5\partial_4\alpha^{(i)}\underline{\partial}_5\partial_2\alpha^{(j)}\underline{\partial}_5\partial_0\alpha^{(k)}\underline{\partial}_3\partial_2\alpha^{(l)}\underline{\partial}_3\partial_0\alpha^{(m)} \rceil,$$

where the constraint $0 < k+l$ in the fifth summation follows since $i+j < M$, and all indices are non-negative.

Now, the first and fifth summations cancel, and we may collect the second with the fourth and the third with the sixth to obtain

$$[\underline{\partial}_5 + \underline{\partial}_3](\omega^{(M)}) =$$
$$\sum_{\substack{i+j+k=M \\ 0\le i,j,k<M}} \lceil \underline{\partial}_5\partial_1\alpha^{(i)}\underline{\partial}_3\partial_1\alpha^{(j)}\underline{\partial}_3\partial_3\alpha^{(k)} \rceil$$

$$- \sum_{\substack{i+j+k+l+m=M \\ 0\le i,j,k,l,m<M}} \lceil \underline{\partial}_5\partial_4\alpha^{(i)}\underline{\partial}_5\partial_2\alpha^{(j)}\underline{\partial}_5\partial_0\alpha^{(k)}\underline{\partial}_3\partial_2\alpha^{(l)}\underline{\partial}_3\partial_0\alpha^{(m)} \rceil.$$

In terms of the associahedron, this last expression is essentially the difference of all composites of total index M (of maps with indices less than M) along the two directed paths around the third and fifth faces. We should also observe (as it will be needed as an analogue of the vanishing condition in the next step) that an essentially identical calculation shows that for $N < M$ we have

$$
\begin{aligned}
0 &= [\underline{\partial}_5 + \underline{\partial}_3](\delta(\alpha^{(N)} + \omega^{(N)})) \\
&= \sum_{\substack{i+j+k=N \\ 0 \le i,j,k \le N}} \lceil \underline{\partial}_5\partial_1\alpha^{(i)}\underline{\partial}_3\partial_1\alpha^{(j)}\underline{\partial}_3\partial_3\alpha^{(k)} \rceil \\
&\quad - \sum_{\substack{i+j+k+l+m=N \\ 0 \le i,j,k,l,m \le N}} \lceil \underline{\partial}_5\partial_4\alpha^{(i)}\underline{\partial}_5\partial_2\alpha^{(j)}\underline{\partial}_5\partial_0\alpha^{(k)}\underline{\partial}_3\partial_2\alpha^{(l)}\underline{\partial}_3\partial_0\alpha^{(m)} \rceil.
\end{aligned}
$$

So we now have

$$
\begin{aligned}
[\underline{\partial}_5 + \underline{\partial}_3 + \underline{\partial}_1](\omega^{(M)}) &= \\
\sum_{\substack{i+j+k=M \\ 0 \le i,j,k < M}} &\lceil \underline{\partial}_5\partial_1\alpha^{(i)}\underline{\partial}_3\partial_1\alpha^{(j)}\underline{\partial}_3\partial_3\alpha^{(k)} \rceil \\
- \sum_{\substack{i+j+k+l+m=M \\ 0 \le i,j,k,l,m < M}} &\lceil \underline{\partial}_5\partial_4\alpha^{(i)}\underline{\partial}_5\partial_2\alpha^{(j)}\underline{\partial}_5\partial_0\alpha^{(k)}\underline{\partial}_3\partial_2\alpha^{(l)}\underline{\partial}_3\partial_0\alpha^{(m)} \rceil \\
+ \sum_{\substack{i+j=M \\ 0 \le i,j < M}} &\lceil \underline{\partial}_1\partial_1\alpha^{(i)}\underline{\partial}_1\partial_3\alpha^{(j)} \rceil \\
- \sum_{\substack{i+j+k=M \\ 0 \le i,j,k < M}} &\lceil \underline{\partial}_1\partial_4\alpha^{(i)}\underline{\partial}_1\partial_2\alpha^{(j)}\underline{\partial}_1\partial_0\alpha^{(k)} \rceil.
\end{aligned}
$$

Now, note that $\underline{\partial}_5\partial_1 = \underline{\partial}_1\partial_4$ and $\underline{\partial}_3\partial_1 = \underline{\partial}_1\partial_2$, while by the naturality of $\alpha^{(0)}$ and its prolongations we have $\lceil\underline{\partial}_3\partial_3\rceil = \lceil\underline{\partial}_4\partial_3\rceil$ and $\lceil\underline{\partial}_1\partial_0\rceil = \lceil\underline{\partial}_0\partial_0\rceil$. Also observe that the $k=0$ terms (for corresponding i,j) of the first and fourth sums are identical, and thus cancel, giving us

$$
[\underline{\partial}_5 + \underline{\partial}_3 + \underline{\partial}_1](\omega^{(M)}) =
$$

$$\sum_{\substack{i+j+k=M \\ 0 \le i,j,k < M \\ 0 < k}} \lceil \underline{\partial}_5 \partial_1 \alpha^{(i)} \underline{\partial}_3 \partial_1 \alpha^{(j)} \underline{\partial}_4 \partial_3 \alpha^{(k)} \rceil$$

$$- \sum_{\substack{i+j+k+l+m=M \\ 0 \le i,j,k,l,m < M}} \lceil \underline{\partial}_5 \partial_4 \alpha^{(i)} \underline{\partial}_5 \partial_2 \alpha^{(j)} \underline{\partial}_5 \partial_0 \alpha^{(k)} \underline{\partial}_3 \partial_2 \alpha^{(l)} \underline{\partial}_3 \partial_0 \alpha^{(m)} \rceil$$

$$+ \sum_{\substack{i+j=M \\ 0 \le i,j < M}} \lceil \underline{\partial}_1 \partial_1 \alpha^{(i)} \underline{\partial}_1 \partial_3 \alpha^{(j)} \rceil$$

$$- \sum_{\substack{i+j+k=M \\ 0 \le i,j,k < M \\ 0 < k}} \lceil \underline{\partial}_5 \partial_1 \alpha^{(i)} \underline{\partial}_3 \partial_1 \alpha^{(j)} \underline{\partial}_0 \partial_0 \alpha^{(k)} \rceil .$$

Again we will use vanishing conditions (the hypothesized one for single face, and the one noted above for the pair of faces) to rewrite the first and last sums, giving

$$[\underline{\partial}_5 + \underline{\partial}_3 + \underline{\partial}_1](\omega^{(M)}) =$$

$$- \sum_{\substack{i+j+k+l=M \\ 0 \le i,j,k,l < M \\ 0 < k,l}} \lceil \underline{\partial}_5 \partial_1 \alpha^{(i)} \underline{\partial}_3 \partial_1 \alpha^{(j)} \underline{\partial}_1 \partial_0 \alpha^{(k)} \underline{\partial}_4 \partial_3 \alpha^{(l)} \rceil$$

$$+ \sum_{\substack{i+j+k=M \\ 0 \le i,j,k < M \\ 0 < k}} \lceil \underline{\partial}_1 \partial_1 \alpha^{(i)} \underline{\partial}_1 \partial_3 \alpha^{(j)} \underline{\partial}_4 \partial_3 \alpha^{(k)} \rceil$$

$$- \sum_{\substack{i+j+k+l+m=M \\ 0 \le i,j,k,l,m < M}} \lceil \underline{\partial}_5 \partial_4 \alpha^{(i)} \underline{\partial}_5 \partial_2 \alpha^{(j)} \underline{\partial}_5 \partial_0 \alpha^{(k)} \underline{\partial}_3 \partial_2 \alpha^{(l)} \underline{\partial}_3 \partial_0 \alpha^{(m)} \rceil$$

$$+ \sum_{\substack{i+j=M \\ 0 \le i,j < M}} \lceil \underline{\partial}_1 \partial_1 \alpha^{(i)} \underline{\partial}_1 \partial_3 \alpha^{(j)} \rceil$$

$$+ \sum_{\substack{i+j+k+l=M \\ 0 \le i,j,k,l < M \\ 0 < k,l}} \lceil \underline{\partial}_5 \partial_1 \alpha^{(i)} \underline{\partial}_3 \partial_1 \alpha^{(j)} \underline{\partial}_3 \partial_3 \alpha^{(k)} \underline{\partial}_0 \partial_0 \alpha^{(k)} \rceil$$

$$- \sum_{\substack{i+j+k+l+m+n=M \\ 0 \le i,j,k,l,m,n < M \\ 0 < n}} \lceil \underline{\partial}_5 \partial_4 \alpha^{(i)} \underline{\partial}_5 \partial_2 \alpha^{(j)} \underline{\partial}_5 \partial_0 \alpha^{(k)} \underline{\partial}_3 \partial_2 \alpha^{(l)} \underline{\partial}_3 \partial_0 \alpha^{(m)} \underline{\partial}_0 \partial_0 \alpha^{(n)} \rceil$$

Applying naturality squares shows that the first and fifth sums are equal and thus cancel. Collecting the second and fourth sums and the third and sixth then gives the desired result. □ □

We now turn to the question of obstructions for fibred and total deformations of monoidal functors, and for deformations of multiplications on monoidal categories (or equivalently, of braided monoidal categories).

Since fibred deformations and deformations of multiplications are special cases of total deformations, defined by restricting the deformation of the target to be trivial or the deformation of the source to be the diagonal deformation induced by the deformation of the target, respectively, it suffices to consider obstructions in the case of total deformations. We begin by giving an explicit formula for these obstructions, and then show that they are closed.

Recall that the appropriate deformation complex for total deformations of a strong monoidal functor (or oplax semigroupal functor)

$$F : \mathcal{C} \to \mathcal{D}, \phi : F(- \otimes -) \Rightarrow F(-) \otimes F(-)$$

is

$$X^\bullet_{\text{total}}(F) = C^\bullet_{\lceil F(p_1) \rceil - \lceil (p_2)_F \bullet \rceil} = X^\bullet(F) \oplus X^{\bullet+1}(\mathcal{C}) \oplus X^{\bullet+1}(\mathcal{D})$$

with coboundary given by

$$\begin{bmatrix} \delta_F & 0 & 0 \\ \lceil F(-) \rceil & -\delta_{\mathcal{C}} & 0 \\ -\lceil (-)_F \bullet \rceil & 0 & -\delta_{\mathcal{D}} \end{bmatrix}.$$

Thus, a cochain will have coboundary which vanishes in each of the second and third coördinates if and only if its second and third coördinates are cocycles in $X^{\bullet+1}(\mathcal{C})$ and $X^{\bullet+1}(\mathcal{D})$, respectively. Similarly, it is easy to see that the obstruction cochain for a total deformation must have as second and third coördinates the obstructions for the deformations of the source and target category, respectively.

Thus, we are left to consider the value of the first coördinate of the obstruction, and the value of the first coördinate of the coboundary. Consider the hexagonal coherence diagram for oplax monoidal functors given in Figure 3.3, with the maps replaced by their deformed versions.

Calculating the difference of the degree n terms of the two directions around the diagram gives

$$\sum_{i+j+k=n} \lceil F(\alpha^{(i)}_{A,B,C} \Phi^{(j)}_{A,B\otimes C} [F(A) \otimes \Phi^{(k)}_{B,C}]] -$$

$$\sum_{i+j+k=n} \lceil \Phi^{(i)}_{A\otimes B,C} [\Phi^{(j)}_{A,B} \otimes F(C)] a^{(k)}_{F(A),F(B),F(C)} \rceil,$$

where α and a are the associators for $\mathcal{C}$ and $\mathcal{D}$, respectively. This must vanish for $n = 1$ for first order total deformations: the vanishing is simply the cocycle condition in $X^\bullet_{\text{total}}(F)$. For a deformation to extend to an Nth order deformation this quantity must vanish for all $n \leq N$, and indeed in addition to the vanishing of the corresponding second and third coördinates, this condition is sufficient. Separating out the terms in which the index $^{(n)}$ occurs, we find that the vanishing conditions are precisely the condition that $[\phi^{(n)}, \alpha^{(n)}, a^{(n)}]$ cobounds $[\Omega^{(n)}, \omega^{(n)}, o^{(n)}]$, where

$$\Omega^{(n)} = \sum_{\substack{i+j+k=n \\ i,j,k < n}} \lceil F(\alpha^{(i)}_{A,B,C} \Phi^{(j)}_{A,B\otimes C} [F(A) \otimes \Phi^{(k)}_{B,C}]]$$

$$- \sum_{\substack{i+j+k=n \\ i,j,k < n}} \lceil \Phi^{(i)}_{A\otimes B,C} [\Phi^{(j)}_{A,B} \otimes F(C)] a^{(k)}_{F(A),F(B),F(C)}$$

and $\omega^{(n)}$ and $o^{(n)}$ are the obstructions to the extension of the deformations of the source and target categories, respectively.

All that remains to show is that the first coördinate of the coboundary of $[\Omega^{(n)}, \omega^{(n)}, o^{(n)}]$ vanishes. We leave the details of the proof to the reader. The method is identical to that applied in the case of the obstructions for deformations of a semigroupal category, except that the associahedron must be replaced with the diagram given in Figure 16.2. In Figure 16.2 we have suppressed all object and arrow labels except for the objects on the inner pentagon which are written with the null infix in place of $\otimes$ to save space.[1] The labels can be recovered by labeling all radial maps with prolongations of Φ and all maps parallel to those between the labeled objects with prolongations of (functorial images of) α. All hexagons are prolongations of the coherence hexagon for semigroupal functors, and all squares, except the diamond-shaped one in the top center, are naturality squares. The diamond is a functoriality square.

---

[1] The diagram of Figure 16.2 was given the name "the Chinese lantern" due to its resemblance to a paper lantern when it is drawn in perspective as the edges of a 3-dimensional polytope with the innermost and outermost pentagons a parallel horizontal faces.

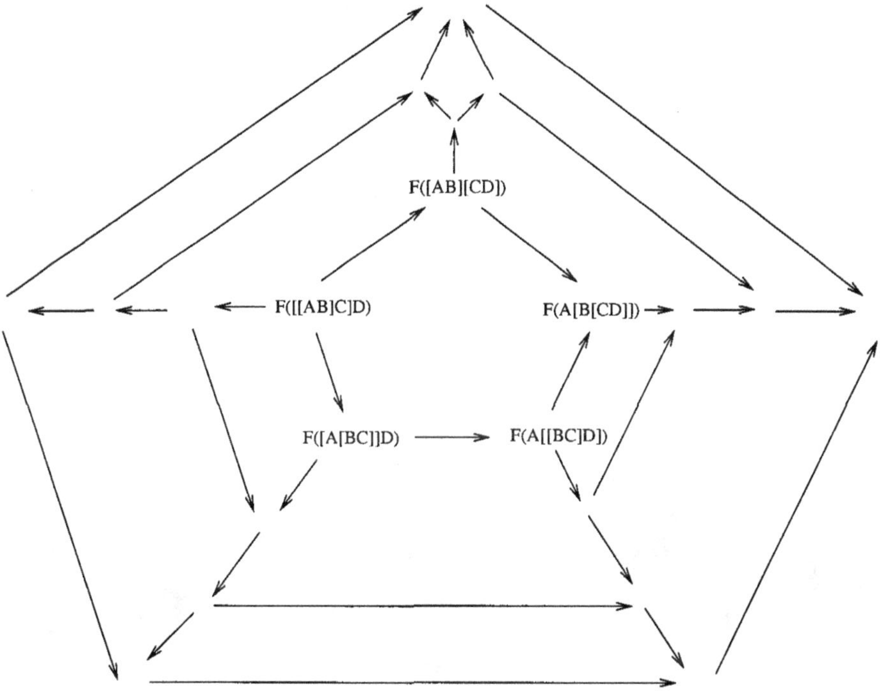

Figure 16.2: The "Chinese Lantern"

Chapter 17

Units

Thus far we have dealt exclusively with deformations of semigroupal categories and functors. We now turn to the question of how deformations of the product structure are related to the presence of a unit object. We will deal first with the question of deformations of monoidal categories as such, then to deformations of monoidal functors. Finally, we will consider the additional condition involving the unit in the definition of a multiplication on a monoidal category.

One reason we have waited this long to consider units is the fact that the condition needed to ensure that a semigroupal deformation is a monoidal deformation is vacuous:

Theorem 17.1 *Every semigroupal deformation of a monoidal category* $(\mathcal{C}, \otimes, I, \alpha, \rho, \lambda)$ *becomes a monoidal category when equipped with unit transformations* $\tilde{\rho}$ *and* $\tilde{\lambda}$ *given by*

$$\tilde{\nu} = \sum_i \nu^{(i)} \epsilon^i$$

$$\tilde{\lambda}_A = \sum_i \lambda_A^{(i)} \epsilon^i$$

$$\tilde{\rho}_A = \sum_i \rho_A^{(i)} \epsilon^i,$$

where

$$\lambda_A^{(n)} = \sum_{i+j=n} \lceil \alpha_{A,I,I}^{(i)}[A \otimes \nu^{(j)}]\rceil\rceil$$

$$\rho_B^{(n)} = \sum_{i+j=n} \lceil \beta_{I,I,B}^{(i)}[\nu^{(j)} \otimes B]\rceil,$$

$\nu^{(i)} : I \to I$ *is any family of maps satisfying* $\nu^{(0)} = \lambda_I = \rho_I$, *and* $\tilde{\alpha}^{-1} = \sum_i \beta^{(i)} \epsilon^i$.

proof: Note that the conditions defining $\tilde{\lambda}$ and $\tilde{\rho}$ can be read off the triangle coherence condition on $\tilde{\alpha}$, $\tilde{\lambda}$, and $\tilde{\rho}$ (see Figure 3.1) in the cases $B = I$ and $A = I$, respectively.

We must verify the triangle condition on $\tilde{\alpha}$, $\tilde{\lambda}$, and $\tilde{\rho}$ in general.

Writing this condition out degree by degree, we find that it is equivalent to

$$\sum_{i+k=N} \lceil [\alpha_{A,I,I}^{(i)} \otimes B][A \otimes \nu^{(k)} \otimes B]\rceil =$$

$$\sum_{i+j+k=N} \lceil \alpha_{A,I,B}^{(i)}[A \otimes \beta_{I,I,B}^{(j)}][A \otimes \nu^k \otimes B]\rceil$$

for all N.

It is thus clear that for any $\tilde{\nu} = \sum_i \nu^{(i)} \epsilon^i$ it suffices to show that

$$\lceil \alpha_{A,I,I}^{(N)} \otimes B\rceil = \sum_{i+j=N} \lceil \alpha_{A,I,B}^{(i)}[A \otimes \beta_{I,I,B}^{(j)}]\rceil$$

for all N. That is,

$$\lceil \tilde{\alpha}_{A,I,I}^{(N)} \otimes B\rceil = \lceil \tilde{\alpha}_{A,I,B}[A \otimes \tilde{\beta}_{I,I,B}]\rceil.$$

Inverting β and writing out term by term, this condition becomes

$$\lceil\alpha_{A,I,B}^{(N)}\rceil - \sum_{i+j=N}\lceil[\alpha_{A,I,I}^{(i)}\otimes B][A\otimes\alpha_{I,I,B}^{(j)}]\rceil = 0.$$

Denote the LHS of this equation by $U_{A,B}^{(N)}$.

Now, any semigroupal deformation satisfies

$$\sum_{i+j=N}\lceil\alpha_{A\otimes B,C,D}^{(i)}\alpha_{A,B,C\otimes D}^{(j)}\rceil -$$
$$\sum_{i+j+k=N}\lceil[\alpha_{A,B,C}^{(i)}\otimes D]\alpha_{A,B\otimes C,D}^{(j)}[A\otimes\alpha_{B,C,D}^{(k)}]\rceil = 0.$$

Specializing to the case where the middle two objects are I (and recalling the effect of $\lceil\ \rceil$), we have

$$\sum_{i+j=N}\lceil\alpha_{A,I,B}^{(i)}\alpha_{A,I,B}^{(j)}\rceil -$$
$$\sum_{i+j+k=N}\lceil[\alpha_{A,I,I}^{(i)}\otimes B]\alpha_{A,I,B}^{(j)}[A\otimes\alpha_{I,I,B}^{(k)}]\rceil = 0.$$

Applying Lemma 3.27 to the last two factors of each summand of the left-hand side allows us to rewrite this as

$$\sum_{i+j=N}\lceil\alpha_{A,I,B}^{(i)}\alpha_{A,I,B}^{(j)}\rceil - \sum_{h+j+k=N}\lceil[\alpha_{A,I,I}^{(h)}\otimes B][A\otimes\alpha_{I,I,B}^{(k)}]\alpha_{A,I,B}^{(j)}\rceil = 0.$$

Rewriting to collect terms according to the last factor gives

$$\sum_{i+j=N}\left\lceil\left[\alpha_{A,I,B}^{(i)} - \sum_{h+k=i}[\alpha_{A,I,I}^{(h)}\otimes B][A\otimes\alpha_{I,I,B}^{(k)}]\right]\alpha_{A,I,B}^{(j)}\right\rceil = 0,$$

but this is simply

$$\sum_{i+j=N}\lceil U_{A,B}^{(i)}\alpha_{A,I,B}^{(j)}\rceil = 0.$$

Now, proceeding by induction, observe that $U_{A,B}^{(0)}$ vanishes by the triangle coherence condition for the undeformed category. Suppose we have shown that $U_{A,B}^{(i)}$ vanishes for all $i < N$. Then $U_{A,B}^{(N)}$ vanishes since the preceding equation reduces to $\lceil U_{A,B}^{(N)} \alpha_{A,I,B}^{(0)} \rceil = 0$, $\alpha_{A,I,B}^{(0)} = \alpha_{A,I,B}$, and all of the padding maps are invertible. $\square$

We now turn to the question of deformations of monoidal functors. In the case of strong monoidal functors, the situation is quite favorable.

Theorem 17.2 *If $(F : C \to D, \Phi, F_o)$ is a strong monoidal functor, then every semigroupal deformation of F extends uniquely to a deformation as a monoidal functor.*

proof: Consider the diagram in Figure 17.1. The maps denoted with tildes ($\tilde{\ }$) are structure maps for some given total deformation of F with chosen monoidal deformation structures for the source and target categories.

Observe that the outside of the diagram commutes by the coherence condition for monoidal functors; the top center square commuted by naturality of $\tilde{a}$ for *any* map $\tilde{F_o}$; the top center triangle commutes by the triangle condition; the bottom center triangle commutes by the triangle condition and the functoriality of F; and the two squares near the bottom commute by the naturality of $\tilde{F}$.

This means that the top left and top right regions (which are functorial images of the two coherence conditions on $\tilde{F_o}$) are the only regions whose commutativity is in doubt. Now, it follows from the naturality and invertibility of F_o and l, that letting

$$\tilde{F_o} = l_{F(I)}[F(I) \otimes F_o^{-1}]\tilde{\Phi}_{I,I}^{-1}F(\tilde{\lambda}_I)\tilde{l}_{F(I)}^{-1}[I \otimes F_o]l_I^{-1}$$

gives a map for which the top right region commutes for any A with $B = I$. It thus follows in this case that the entire diagram commutes when beginning at the bottom left. By the invertibility of $\tilde{\Phi}$, it follows that the top left region commutes in this case. But $F(I)$ is isomorphic to I, and thus $-\otimes F(I)$ is faithful. Therefore for any A, the given map $\tilde{F_o}$

satisfies the coherence condition for monoidal functors involving right unit transformations.

The same argument shows that if $\tilde{F}_o$ is any map which makes the top left region commute in the case of $A = I$, it makes the top right region commute for any B, and thus satisfies the coherence condition for monoidal functors involving left unit transformations. $\square$

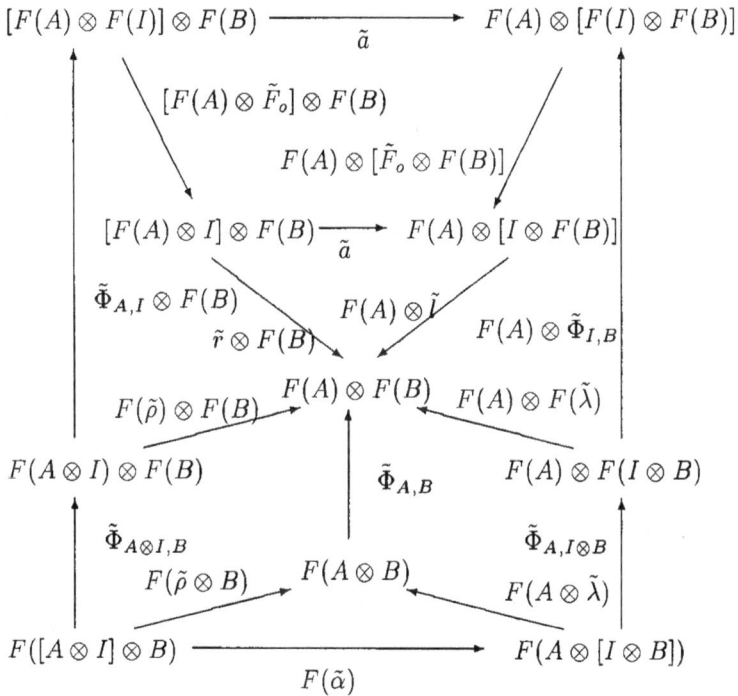

Figure 17.1: Diagram Relating Unit Conditions for Strong Monoidal Functors

The final conditions involving units which must be considered are the conditions in the definition of a multiplication on a monoidal category. This condition, however, is trivially satisfied by any deformation, since the isomorphism giving the structure of the given multiplication still provides the necessary structure after deformation.

Chapter 18

Extrinsic Deformations of Monoidal Categories

Thus far we have been considering what might be called "intrinsic" deformations of monoidal categories, monoidal functors and braided monoidal categories. We now wish to turn to a more general type of deformation: deformations which take place in the context of a larger ambient category.

A priori we can consider embeddings of the monoidal category $\mathcal{C}$ into an arbitrary category. Such an approach faces substantial difficulties, since the natural transformations must be natural not only with respect to the maps in $\mathcal{C}$, but also with respect to their own components. We therefore consider the case of a strict monoidal inclusion.

We begin with the observation:

Proposition 18.1 *Let $\iota : \mathcal{C} \to \mathcal{X}$ be a strict monoidal (resp. braided, ribbon) inclusion. Let $\mathbf{full}(\mathcal{C})$ be the full-subcategory generated by $\mathcal{C}$. Then $\mathbf{full}(\mathcal{C})$ is a monoidal (resp. braided, ribbon) category when equipped with the same structure maps as $\mathcal{C}$ (that is, the restriction of the structure maps of $\mathcal{X}$).*

We can then make

Definition 18.2 *An* extrinsic deformation *of a monoidal (resp. braided) category C is a strict monoidal (resp. braided) inclusion $\iota : C \to \mathcal{X}$, together with a deformation of $\mathcal{X}$. Two extrinsic deformations are* equal *if the restrictions of the deformations of $\mathcal{X}$ to the full-subcategory generated by C are equal. Two extrinsic deformations are* equivalent *if the restrictions of the deformations of $\mathcal{X}$ to the full-subcategory generated by C are monoidally (resp. braided) equivalent.*

What is important for our approach to Vassiliev theory, which we will begin in the next chapter, is the fact that extrinsic deformations of a monoidal (resp. braided) category can be organized into a category.

Definition 18.3 *Given an R-linear monoidal (resp. braided) category C, the* category of nth order extrinsic deformations of C, *which we denote* $\mathbf{Extr}_n(C)$, *has as objects all n^{th} order extrinsic deformations of C. Given two extrinsic deformations*

$$(\iota : C \to \mathcal{X}, \psi^{(0)}, \ldots, \psi^{(n)})$$

and

$$(\zeta : C \to \mathcal{Z}, \varphi^{(0)}, \ldots \varphi^{(n)})$$

(where $\psi(k)$ and $\varphi(k)$ are the cocycles in the appropriate deformation complex), a map from $(\iota, \vec{\psi})$ to $(\zeta, \vec{\varphi})$ is an $R[\epsilon]/ < \epsilon^n >$-linear strict monoidal (resp. braided) functor

$$\Phi : \mathbf{full}(\iota(C))^{(n)} \to \mathbf{full}(\zeta(C))^{(n)} \ ,$$

where both the source and target are equipped with the deformed monoidal (resp. braided) structure, which reduces modulo ϵ to a strict monoidal functor Φ_0 satisfying

$$\Phi_0(\iota) = \zeta.$$

$\mathbf{Extr}_\infty(C)$ *is defined similarly.*

What is important to note here is that $\mathbf{Extr}_n(\mathcal{C})$ and $\mathbf{Extr}_\infty(\mathcal{C})$ have initial objects. This follows from a general argument of a type we have already seen: whether monoidal or braided, once a ground ring and a category $\mathcal{C}$ are fixed, the extrinsic deformations turn out to be models of an essentially algebraic theory (possibly with infinitely many constants and equations to capture the structure of the ring and $\mathcal{C}$, but this is of no consequence to the argument), while our maps are simply homomorphisms of models of this theory.

Observe that intrinsic deformations are precisely those extrinsic deformations for which $\iota : \mathcal{C} \to \mathcal{X} = Id_{\mathcal{C}} : \mathcal{C} \to \mathcal{C}$. Likewise, extrinsic deformations for a fixed $\iota : \mathcal{C} \to \mathcal{X}$ are classified by the appropriate deformation complex for the monoidal (resp. braided) category $\mathbf{full}(\mathcal{C})$.

The initial object is then a "free" deformation of $\mathcal{C}$. In the next section we will see that the free braided deformation of the free rigid symmetric R-linear monoidal category is related in an important way to the theory of R-linear Vassiliev invariants.

Chapter 19

Vassiliev Invariants, Framed and Unframed

It is usual to discuss Vassiliev theory in terms of unframed oriented knots and links (cf. [8], [51], [55]). We will, however, for the most part remain in the setting most natural for functorial invariants (and incidentally most closely connected to 3- and 4-manifold topology), that of framed links. Until we reduce the subject to its combinatorial content, we will here switch briefly from the PL to the smooth setting.

Recall our discussion of Goryunov's approach to framed links as equivalence classes of embeddings of ribbon neighborhoods in Chapter 8. In order to discuss Vassiliev theory, we now drop the requirement that the mappings be embeddings.

The space of possibly-singular framed links is the space of all C^∞ mappings of ribbon neighborhoods of disjoint unions of circles (topologized with the topology of uniform convergence on compacta of the map and all its derivatives) modulo this equivalence relation. Fix a pair A, B of equivalence classes of maps from ribbon neighborhoods of germ neighborhoods of signed sets of points to a germ neighborhood of $\mathbb{I}^2$. The space of possibly-singular framed tangles with source A and target B is the space of all C^∞ mappings of ribbon neighborhoods of disjoint unions of circle and intervals which intersect the boundary of $\mathbb{I}^3$ only in boundary points of the underlying ribbon neighborhood, which are mapped by a map equivalent to A to the top face and a map equivalent to B to the bottom face. We denote the former by Ω_f^{lk} and the latter by $\Omega_f(A, B)$. We will either of these spaces by Ω_f when no confusion is possible or when it is a matter of indifference which is meant. The subspace of equivalence classes of mappings of neighborhoods of $S_1 \cup \ldots \cup S_k$ is the space of possibly-singular framed links of k components, and will be denoted $\Omega_f(k)$. Now consider the subspaces $\mathcal{O}_f^{\mathrm{lk}} \subset \Omega_f^{\mathrm{lk}}$ and $\mathcal{O}_f(A, B) \subset \Omega_f(A, B)$ of all (equivalence classes of) mappings such that Tg is an embedding.

Definition 19.1 *A* (non-singular) framed link (resp. framed tangle) *is a connected component of* $\mathcal{O}_f$. *A connected component of* $\mathcal{O}_f(k) = \mathcal{O}_f \cap \Omega_f(k)$ *is a* (non-singular) framed link of k components. *A connected component of* $\mathcal{O}(A, B)$ *is a* (non-singular) smooth tangle with source A and target B.

Observe that this agrees with the smooth version of our previous definition.

First, note that we will often drop the adjective "non-singular" to match the usual usage in knot theory (as in earlier chapters). Second, the designation of these maps as non-singular implicitly identifies the element of $\Omega_f \setminus \mathcal{O}_f$ as *singular*. We denote the discriminant locus $\Omega_f \setminus \mathcal{O}_f$ by Σ_f. As observed in Goryunov [26], the discriminant is the union of two hypersurfaces (interpreted in the appropriate infinite dimensional sense), one on which the framing degenerates, denoted Σ_f', and one on which the disjoint union of circles is not embedded, denoted Σ_f''.

The minimal degenerations of each type are illustrated in Figure 19.1. In the top part of Figure 19.1 the diagonal line indicates the framing curve (the edge of the ribbon obtained by applying the exponential map for the standard metric to the framing vectors), while in the bottom part we suppress indication of the framing. We denote the subspace of Σ_f in which there are exactly n degenerations of either type by $\Sigma_{f,n}$, so $\mathcal{O}_f = \Sigma_{f,0}$. Then letting $\Omega_{f,i} = \Omega_f \setminus (\cup_{j=0}^{i-1} \Sigma_{f,j})$, Ω_f becomes a stratified space. (This is true for both Ω_f^{lk} and $\Omega_f(A, B)$, and by intersecting with $\Omega_f(k)$ we obtain a stratification of this space as well.)

As in Goryunov [26], we coörient the finite codimensional strata of Σ_f by the local prescriptions given in Figure 19.1, and give a Vassiliev-type prescription for the extension of invariants of framed knots (links or tangles), to singular framed knots (links or tangles) as in Figure 19.2.

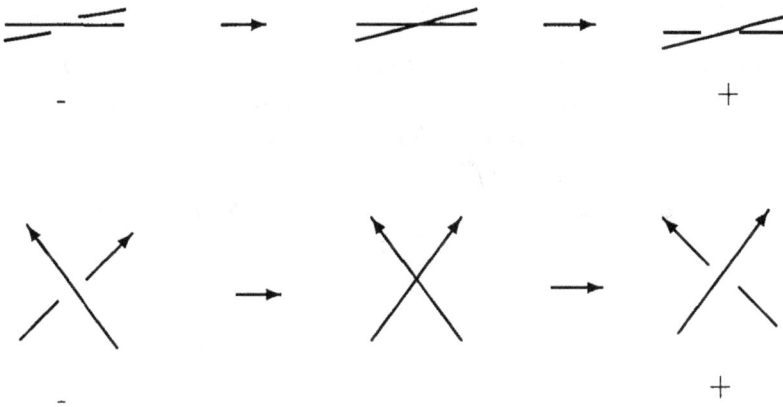

Figure 19.1: Coörienting the Finite Codimensional Strata

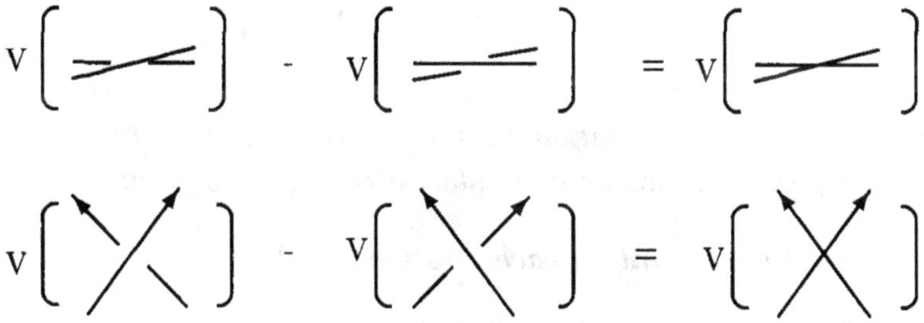

Figure 19.2: Vassiliev-type Extension Formulae

Definition 19.2 *An R-valued Vassiliev invariant of framed links (resp. framed links of k components, framed tangles) is a locally constant R-valued function on $\mathcal{O}_f$ (resp. $\mathcal{O}_f(k)$) whose extension, according to the prescription of Figure 19.2, vanishes on $\Sigma_{f,n+1}$ (resp. $\Sigma_{f,n+1}(k)$) for some n, in which case the invariant is said to be of type $\leq n$. The invariant is of type N when N is the minimal such n.*

It is then clear that

Proposition 19.3 *R-valued Vassiliev invariants of framed links (resp. of framed links of k components) form an R-module $\mathcal{V}$ (resp. $\mathcal{V}(k)$) under pointwise addition and scalar multiplication. Similarly, the R-valued Vassiliev invariants of type $\leq n$ form an R-module $\mathcal{V}_n$ (resp. $\mathcal{V}_n(k)$) under the same operations, and the natural inclusions $\mathcal{V}_{n-1} \hookrightarrow \mathcal{V}_n$ (resp. $\mathcal{V}_{n-1}(k) \hookrightarrow \mathcal{V}_n(k)$) make $\mathcal{V}$ (resp. $\mathcal{V}(k)$) into a filtered R-module.*

Observe also that as we have defined them, the possibly singular framed tangles form a stratified category $\Omega_{\mathbf{f}}$, whose underlying ordinary category has a monoidal structure, which is, in fact, given by stratified functors. The infinite codimension part of the hom-stratified-spaces should be discarded to obtain a stratified category $\Omega_{\mathbf{f}}^{\text{fin}}$ whose hom-stratified-spaces are $\Omega_f(A, B)^{\text{fin}}$. Observe that

$$X^{\text{fin}} \otimes Y^{\text{fin}} = (X \otimes Y)^{\text{fin}} .$$

Definition 19.4 *A Vassiliev invariant of framed tangles over R is a monoidal functor V from the underlying category of $\Omega_{\mathbf{f}}^{\text{fin}}$ to the underlying category of an R-linear monoidal category, satisfying:*

1. *V is locally constant on each stratum $\Sigma_f(A, B)_i$.*

2. *If $T \in \Sigma_f(A, B)_{i+1}$ and $T_+, T_- \in \Sigma_f(A, B)_i$ are related to T by the removal of one degeneracy as in Figure 19.1, with the sign indicating which side of the $i + 1^{st}$ stratum relative to the coorientation they lie on, then $V(T) = V(T_+) - V(T_-)$.*

3. *There exists an n such that for all A, B and all $T \in \Sigma_f(A,B)_m$ for $m > n$, $V(T) = 0$.*

We say V is of type $\leq n$ in this case, and of type N where N is the minimal such n.

The connection of Vassiliev theory to the deformation theory of braided monoidal categories is quite direct. In fact, we have:

Theorem 19.5 *Let C be any K-linear rigid symmetric monoidal category, and let $\tilde{C}$ be any n^{th} order tortile deformation of C. For any object X of $\tilde{C}$, let V_X denote the functor from $\mathcal{FT}$ to $\tilde{C}$ induced by Shum's Coherence Theorem. Then V_X restricted to $End(I)$, regarded as the set of framed links, is a $K[\epsilon]/<\epsilon^{n+1}>$-valued Vassiliev invariant of type $\leq n$, and is, moreover, multiplicative under disjoint union.*

From this will follow, almost as a corollary,

Theorem 19.6 *Let C be any K-linear rigid symmetric monoidal category, and let $\tilde{C}$ be any n^{th} order tortile deformation of C (resp. formal series deformation of C). For any object X of $\tilde{C}$, let V_X denote the functor from $\mathcal{FT}$ to $\tilde{C}$ induced by Shum's Coherence Theorem, and let $V_{X,k}$ denote the K-valued framed link invariant which assigns to any link the coefficient of ϵ^k, for $k = 0, ...n$, (resp. for $k \in \mathbb{N}$). Then $V_{X,k}$ is a Vassiliev invariant of type $\leq k$.*

The proof of Theorem 19.5 is quite simple and similar to previous proofs of similar results:

proof of Theorem 19.5: The key is to observe that the bilinearity of composition in $\tilde{C}$ allows us to use the Vassiliev prescription to extend the functor V_X from $\mathcal{FT}$ to a larger category of singular framed tangles, $\widetilde{\mathcal{FT}}$, whose maps are isotopy classes of framed tangles with finitely many degeneracies of either of the two basic types.

Consider a singular framed link with $n + 1$ degeneracies (of either type). Now, we can represent the framed link as a composition of singular framed tangles, each of which has at most one degeneracy, crossing, framing twist, or extremum. The value of the extended functor on such a tangle with a degeneracy of the first type (framing degeneracy) is a monoidal product of identity maps with $\theta_X - \theta_X^{-1}$ (or its dual), while the value on such a tangle with a degeneracy of the second type is a monoidal product of identity maps with $\sigma_{X,X} - \sigma_{X,X}^{-1}$.

Now, observe that

$$\theta_X - \theta_X^{-1} \in Hom_{\mathcal{C}}(X, X) \otimes <\epsilon>$$

and

$$\sigma_{X,X} - \sigma_{X,X}^{-1} \in Hom_{\mathcal{C}}(X \otimes X, X \otimes X) \otimes <\epsilon>.$$

It follows from the bilinearity of composition in $\tilde{\mathcal{C}}$ that the composite representing the singular framed link as an element of $End_{\tilde{\mathcal{C}}}(I)$ lies in $End_{\mathcal{C}}(I) \otimes <\epsilon^{n+1}> = 0$, thus showing V_X to be Vassiliev of type $\leq n$. Multiplicativity follows from functoriality. $\square$

Theorem 19.6 follows from the Theorem 19.5 and the following lemma, the proof of which is a trivial exercise:

Lemma 19.7 *Let A and B be abelian groups. If V is an A-valued Vassiliev invariant, and $f : A \to B$ is a linear map, then $f(V)$ is a B-valued Vassiliev invariant.*

The connection between Vassiliev theory and deformation theory is even more intimate, as we will see upon further examination. We will not pursue Vassiliev's original cohomological approach (cf. [55]) to the construction of such invariants, although our observations above and Definition 19.4 show that it is possible to pursue this approach in the tangle-theoretic setting. Rather, we will follow the combinatorial approach taken by such authors as Birman and Lin [8], Bar-Natan [6, 5, 7], and the related constructions involving iterated integrals proposed by Kontsevich [37] on the basis of the constructions originally given by Drinfel'd [18, 19].

The key to the combinatorial approach is the observation that the Vassiliev extension prescription and the vanishing condition almost reduce the description of type n Vassiliev invariants to the description of their values on $\Sigma_{f,n}$. "Almost", because there is an ambiguity of a type $n-1$ invariant. More precisely,

Lemma 19.8 *If V_1, V_2 are Vassiliev invariants of type n (that is, vanishing on all (singular framed) knots, links or tangles with more than n degeneracies) and for any $T \in \Sigma_{f,n}$ we have $V_1(T) = V_2(T)$ (in the case of tangles, for all $\Sigma_f(A, B)_n$), then $V_1 - V_2$ is a Vassiliev invariant of type $n-1$. Conversely, if two Vassiliev invariants of type n differ by a Vassiliev invariant of type $n-1$, then they have the same values on all knots, links, or tangles with n degeneracies.*

The proof is a trivial exercise. (Observe that in the tangle case, there is an implied universal quantification over pairs of objects, and the equality implies that the two functors agree on objects.)

It is thus important to have a convenient description for the values of type n invariants on knots, links and tangles with n degeneracies. The usual one, used in, for example, Birman and Lins [8] and Bar-Natan [6], depends upon the following:

Lemma 19.9 *If V is a Vassiliev invariant of type n, and T_1 and T_2 are two (singular framed) knots, links, or tangles with n degeneracies, and differing only by a sequence of crossing changes, then $V(T_1) = V(T_2)$.*

Again, the proof is a trivial exercise.

We may consider two (singular framed) knots, links, or tangles with n degeneracies to be n-equivalent if they are related by a sequence of ambient isotopies rel boundary and crossing changes – including changes to the framing by an even number of twists.

To specify an n-equivalence class, it thus suffices to specify the n degeneracies. This is usually done diagrammatically by indicating the pair of points to be identified by joining them with chords (usually drawn as dashed lines to distinguish them from arcs of the knot, link, or tangle, cf. Birman and Lins [8] and Bar-Natan [6]). We will specify framing degeneracies by placing a large dot or "bead" on the arc of the tangle to indicate the location of the degeneracy.

Thus, an n-equivalence class of singular framed knots, links, or tangles with n degeneracies can be specified by a "chord-and-bead" diagram. (In the unframed case, one uses only chords, thus obtaining the chord diagrams of Birman-Lins [8] and Bar-Natan [6].) The value of a type n Vassiliev invariant on singular knots, links, or tangles is thus a function on the set of n-equivalence classes (or of n-chord-and-bead diagrams). However, because of the fact that it arose from an ambient isotopy invariant by the Vassiliev extension prescription at degeneracies, it cannot be an arbitrary function. Rather, it must satisfy identities which are the reflection of the Reidemeister moves (or equivalently, of the relations in the category of framed tangles) under the extension prescription:

Definition 19.10 *A framed weight system of degree n for framed knots (resp. links of c components) is a function on the set of one- (resp. c-) loop chord-and-bead diagrams to R (for some abelian group R, usually taken to be the additive group of a commutative ring) so that the values satisfy the 4T, side change, bead and bead-slide relations of Figure 19.3.*

Here, as always, the standard convention of skein theory is followed: imposing a relation on parts of diagrams means imposing it on all extensions to complete diagrams identical outside the region depicted.

Thus, the dual of the module of framed weight systems of degree n is the quotient of the R-module obtained from the free R-module on the basis of all n-chord-and-bead diagrams by quotienting by all instances of the relations of Figure 19.3.

In Figure 19.4 we show a basis for the dual module to degree 3 framed weight systems.

To deal with tangles, we consider the symmetric compact closed category of oriented immersed planar diagrams equipped with beads and chords, subject to the relations of Figure 19.3 and Figure 19.5, which we denote by **FrVasTang**$_R$. In Figure 19.5 each relation should

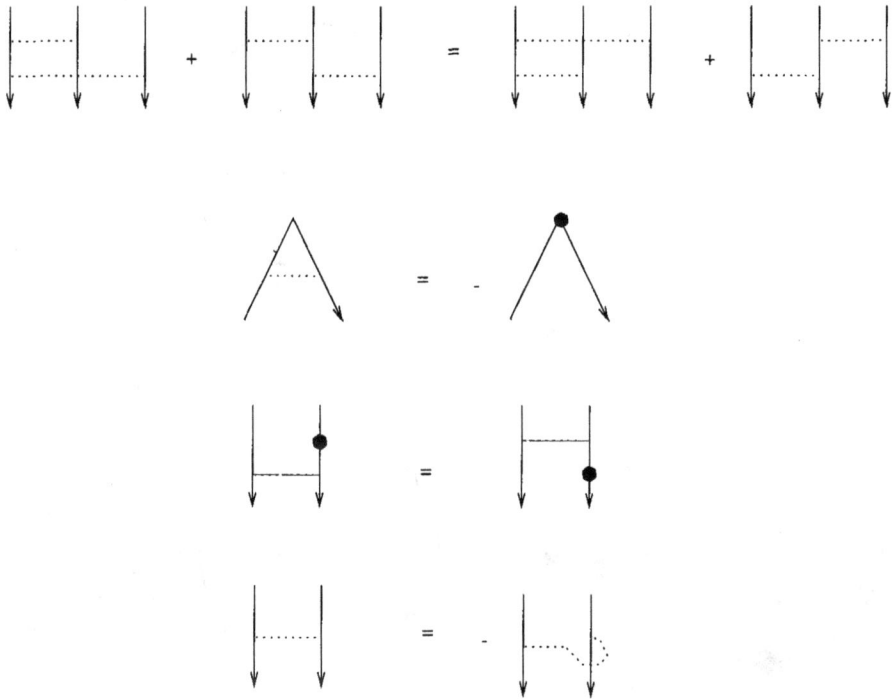

Figure 19.3: Relations on Chord and Bead Diagrams

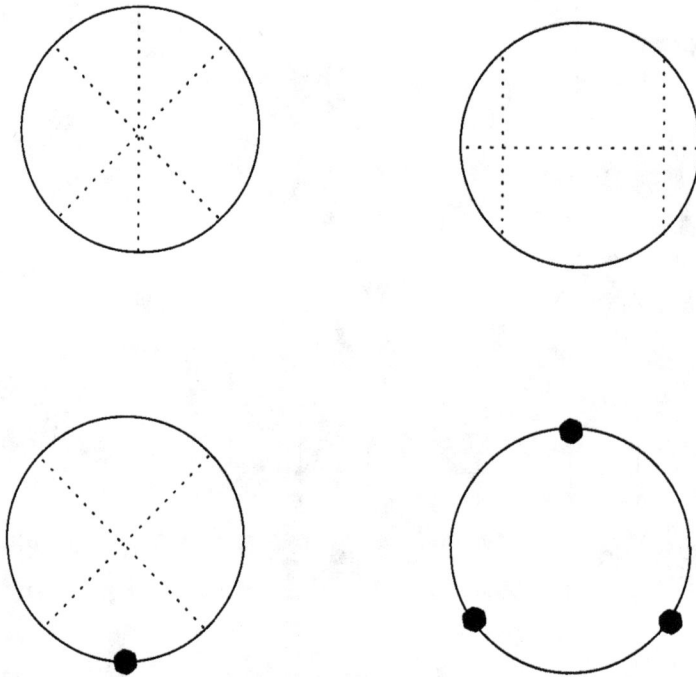

Figure 19.4: A Dual Basis for Degree 3 Framed Weight Systems for Knots

be taken with all possible orientations on the strands. The unframed analog **VasTang**$_R$ is the quotient obtained by setting the bead equal to 0.

The appropriate analog of weight systems for tangles is then a monoidal functor from **FrVasTang**$_R$ or **VasTang**$_R$ to R-mod.

Now, observe that by Kelly and Laplaza's coherence theorem [34] (the proof of which may be extracted by simplifying Shum's proof which it inspired) a category monoidally equivalent to the free symmetric compact closed category $F_{scc}(1)$ is monoidally included in **FrVasTang**$_R$ (resp. **VasTang**$_R$) as the subcategory of diagrams with neither chords nor beads. Isomorphism classes of objects may be identified with words of 1's and 1*'s (or +1's and −1's).

It is then possible to consider the structure of the extrinsic deformations of $R[F_{scc}(1)]$ in **FrVasTang**$_R$ or **VasTang**$_R$. In the case of $R = K$ for $\mathbb{Q} \subset K \subset \mathbb{C}$, this structure turns out to have been extensively investigated by Kontsevich [37], Bar-Natan [6, 5, 7] and others building on the work of Drinfel'd [18, 19].

FrVasTang$_R$ is equipped with natural transformations derived from the chord and bead:

Proposition 19.11 *The family of maps given by letting* $\kappa_{A,B}$ *be the sum of all 1-chord diagrams with a horizontal chord beginning on a strand of A and ending on a strand of B is a natural endormorphism of* $\otimes$. *The family of maps given by letting* τ_A *be the linear combination of all 1-chord-or-bead diagrams with either a bead on a single strand of A or a chord joining a pair of strands of A, with each summand involving a bead having coefficient of 1 and each summand involving a chord having coefficient* −2, *is a natural endomorphism of* $Id_{\mathbf{FrVasTang}_R}$ ·

sketch of proof: We leave the complete check to the reader after observing that for categories described by generators under composition (and relations), it suffices to check naturality at each generating map. Thus, in the present context it suffices to check for all monoidal prolongations of the crossing (twist), cup, cap, chord and bead. For example, the naturality of $\kappa_{A,B}$ at a map consisting of a chord joining two strands

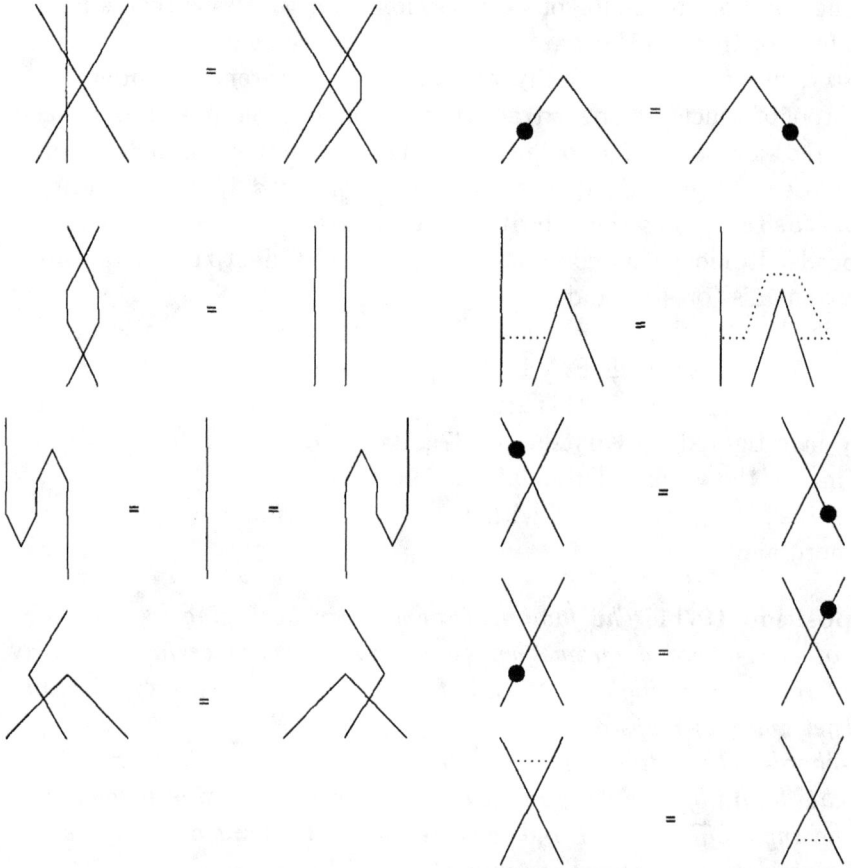

Figure 19.5: Relations on Bead-and-Chord Tangle Diagrams

of A (resp. B) follows immediately from the 4T relation (and the functoriality properties of $\otimes$ which allow chords not on the same strand to be commuted). In both cases naturality with respect to prolongations of the crossing and the bead are quite easy. For the cup and cap, one must use the fourth property of Figure 19.3. $\square$

Observe that this natural transformation for each fixed B (resp. fixed A) satisfies a property analogous to principal-ness of elements in a coalgebra:

Definition 19.12 *A paracoherent natural transformation* $f_{A_1,...,A_n}$ *between structural functors of a symmetric monoidal category is* multi-principal *if it satisfies*

$$\lceil f_{A_1,...,A_k \otimes A_l,...,A_n} \rceil = \lceil f_{A_1,...,A_k,...,A_n} \otimes A_l \rceil + \lceil A_k \otimes f_{A_1,...,A_l,...,A_n} \rceil.$$

It is then easy to prove:

Proposition 19.13 κ, *as defined above, is multiprincipal.*

Of course, this gives a large family of natural endomorphisms of the various iterates of $\otimes$ by composing various monoidal prolongations of the two just given.

We can now restate the main result of [18] as a deformation theoretic statement:

Theorem 19.14 *Let K with $\mathbb{Q} \subset K \subset \mathbb{C}$ be a field. The natural transformations* $\phi^{(1)} = \lceil 1 \otimes \kappa_{A,B} tw_{A,B} \otimes 1 \rceil$ *and* $\alpha^{(1)} = 0 :^3 \otimes \Rightarrow \otimes^3$ *give the first order term of a braided deformation of* **VasTang**$_K$. *Moreover, all obstructions vanish and the higher terms may be chosen so that*

$$\phi^{(n)} = \frac{1}{n!} \lceil 1 \otimes \kappa_{A,B}^n tw_{A,B} \otimes 1 \rceil.$$

The uniqueness up-to-equivalence in the main theorem of [18], turns out to be an immediate consequence of this deformation theoretic statement: if two different cocycles have the same coboundary (in this context, a particular obstruction cocycle whose vanishing both cocycles witness), they are necessarily cohomologous, and the cocycle which cobounds their difference is then the next term in the equivalence between the deformations. Likewise, by the same reasoning, it follows that the higher terms need not be chosen to be the exponential terms suggested by Drinfel'd [18], though any choice will be cohomologous to these.

This result, however, is very much dependent upon the fact that the coefficients lie in a field of characteristic 0. In the next chapter we examine the special case of characteristic 2.

This formulation of the universal Vassiliev invariant in the context of categorical deformation theory makes clear that Vassiliev theory lives quite comfortably within the bounds of functorial knot theory. Thus, the limitations on the topological power of the knot polynomials of Jones, HOMFLY, and Kauffman [29, 42, 32] (e.g. the inability to distinguish flypes) are either not endemic to functorial invariants *per se* or are shared by Vassiliev invariants.

Another consequence of this result is the existence of a non-trivial deformation of **VasTang$_K$** as a monoidal category. Calculating the second order term of the deformation of Theorem 19.14, we find that

$$\alpha^{(2)}_{A,B,C} = \frac{1}{2}[\kappa_{A,B} \otimes C][A \otimes \kappa_{B,C}] - \frac{1}{2}[A \otimes \kappa_{B,C}][\kappa_{A,B} \otimes C].$$

Since the first order term in Drinfel'd's construction was zero, this natural transformation is itself a 3-cocycle, but

Proposition 19.15 $\alpha^{(2)}$, *as given above, represents a non-trival cohomology class.*

proof: The hom-spaces of **VasTang$_K$** are naturally graded by the number of chords in a summand. It thus suffices to show that $\alpha^{(2)}$ is not the coboundary of a natural transformation with two chords.

A moment's thought shows that all such natural transformations from $\otimes$ to itself are scalar multiples of κ^2. It thus suffices to show that $\alpha^{(2)}$ is not a multiple of $\delta(\kappa^2)$.

Computing, we find that

$$\delta(\kappa^2)_{A,B,C} =$$
$$-\lceil[\kappa_{A,C} \otimes B][A \otimes \kappa_{B,C}]\rceil$$
$$-\lceil[A \otimes \kappa_{B,C}][\kappa_{A,C} \otimes B]\rceil$$
$$+\lceil[\kappa_{A,C} \otimes B][\kappa_{A,B} \otimes C]\rceil$$
$$+\lceil[\kappa_{A,B} \otimes C][\kappa_{A,C} \otimes B]\rceil.$$

Using the naturality and multiprincipal properties of κ (the 4T relation and definition), we can rewrite this as

$$\delta(\kappa^2)_{A,B,C} =$$
$$-4\alpha^{(2)}_{A,B,C}$$
$$-2\lceil[A \otimes \kappa_{B,C}][\kappa_{A,C} \otimes B]\rceil$$
$$+2\lceil[\kappa_{A,B} \otimes C][\kappa_{A,C} \otimes B]\rceil.$$

Now, the only relations on natural transformations represented by linear combinations of 2-chord diagrams are instances of the 4T relation. It thus follows that

$$\{\lceil[A \otimes \kappa_{B,C}][\kappa_{A,C} \otimes B]\rceil], \lceil[\kappa_{A,B} \otimes C][\kappa_{A,C} \otimes B]\rceil],$$

$$[\kappa_{A,B} \otimes C][A \otimes \kappa_{B,C}], [A \otimes \kappa_{B,C}][\kappa_{A,B} \otimes C]\}$$

forms a basis for the subspace of $X^3(\mathbf{VasTang_K})$ of natural transformations represented by linear combinations of 2-chord diagrams. Thus, $\alpha^{(2)}$ is not a coboundary. $\square$

Since our deformation theoretic degree need not correspond to the number of chords, we can take $\alpha^{(2)}$ as the first order term in a deformation of the monoidal structure, thus also giving our first example of a non-trivial deformation of a monoidal category as such.

This deformation-theoretic approach also gives a more satisfactory explanation to the "fudge factor" needed to get invariance of the universal Vassiliev invariant under moves of type $\Delta.\pi.1$. Observe that the defining condition for dual objects involves equations containing the associator. Once the category has been deformed, the old structure maps for the dual no longer satisfy the required equation with respect to the new associator. One can choose either ϵ or η to remain unchanged, but the other must be changed to restore the dual structure. A simple calculation shows that if ϵ is unchanged, η must be replaced with $\eta[X^* \otimes \Upsilon^{-1}]$, where Υ is the map $\rho^{-1}[X \otimes \eta]\tilde{a}^{-1}[\epsilon \otimes X]\lambda$, in order for the first of the defining relations to hold. The second follows by symmetry.

Now, observe that in the presence of a "graphical Schur's Lemma" this correction to each occurence of η becomes precisely the identity map on X divided by the trace of Υ. Thus, the correction given in Bar-Natan [6] of dividing by a power of the Kontsevich integral of the unknot with no crossings and two maxima (and two minima) is seen to actually arise from Schur's Lemma, the preservation of a tortile structure and the resulting functorial invariant.

Schur's Lemma is, of course, a consequence of the simplicity of the object of the category corresponding to the downward strand, and thus, in the case of the standard construction from representations of a Lie algebra, of the semi-simplicity of the Lie algebra. Vogel [56] has recently constructed Vassiliev invariants of knots which do not arise from any semi-simple Lie algebra.

Thus, the construction of the universal functorial Vassiliev link invariant must be done in the absence of Schur's Lemma, and the "correction" which arises from the need to modify η (or ϵ) to preserve the tortile structure under deformation must be made by replacing η, that is simple minima in the diagram, by a linear combination of terms which are given precisely by the map of which Bar-Natan's correction factor is the categorical trace.

Now, all categories corresponding to the weight systems constructed from semisimple Lie algebras are semisimple categories, and therefore satisfy Schur's lemma. Thus, the corresponding Vassiliev invariants are specializations of Bar-Natan's corrected Kontevich integral, while the invariants of Vogel [56] will not arise from Bar-Natan's correction, but only from the inclusion of corrections at each minimum of the link.

Chapter 20

Vassiliev Theory in Characteristic 2

The following suffices to establish the different character of deformation theory over fields of characteristic 2, and also provides us with our first example of a non-vanishing obstruction.

Theorem 20.1 *Let k be a field of characteristic 2. Then the natural transformations $\phi^{(1)} = \lceil 1 \otimes \kappa_{A,B} tw_{A,B} \otimes 1 \rceil$ and $\alpha^{(1)} = 0 : {}^3\!\otimes \Rightarrow \otimes^3$ give the first order term of a braided deformation of $\mathbf{VasTang}_k$ for which the second order obstruction does not vanish.*

proof: We proceed by contradiction. Assume the contrary. Now, observe that the homspaces and space of natural tranformations are naturally graded by the number of chords. Observe also that the coboundary operator preserves the number of chords. The obstruction is easily seen to be a linear combination of diagrams with two chords. Now, by Theorem 12.13, any multiplication is equivalent to a multiplication of the form $\lceil 1 \otimes \sigma \otimes 1 \rceil$ for a braiding σ. We may thus assume without loss of generality that the hypothesized deformation term which cobounds the obstruction has first coördinate (deforming the structure map of the multiplication) of the form $a \lceil 1_A \otimes \kappa_{B,C}^2 tw_{B,C} \otimes 1_D \rceil$ for some $a \in k$, with the associated braiding given by $\sigma_{B,C} = [Id_{B \otimes C} + \kappa_{B,C}\epsilon + a\kappa_{B,C}^2] tw_{B,C}$.

Now, the second coördinate (deforming the associator) is necessarily of the form

$$\alpha^{(2)} = b\lceil \kappa_{A,B}^2 \rceil + c\lceil \kappa_{A,B} \rceil \lceil \kappa_{A,C} \rceil + d\lceil \kappa_{A,B} \rceil \lceil \kappa_{B,C} \rceil + e\lceil \kappa_{A,C}^2 \rceil$$
$$+ f\lceil \kappa_{A,C} \rceil \lceil \kappa_{A,B} \rceil + g\lceil \kappa_{B,C} \rceil \lceil \kappa_{A,C} \rceil + h\lceil \kappa_{B,C}^2 \rceil.$$

Observe that by the 4T relation, the two remaining 2-chord diagrams are linear combinations of those represented in this expression.

The deformed associator $\tilde{\alpha} = \alpha^{(0)} + alpha^{(2)}\epsilon^2$ and the deformed braiding $\tilde{\sigma} = [Id+\kappa+a\kappa^2]tw$ must satisfy the two hexagons of Definition 5.1.

Calculating, we find that

$$[\tilde{\sigma}_{A,B} \otimes C]\tilde{\alpha}_{B,A,C}[B \otimes \tilde{\sigma}_{A,C}] = L_{A,B,C}tw_{A,B\otimes C}\epsilon^2 + \text{lower order terms}$$

and

$$\tilde{\alpha}_{A,B,C}\tilde{\sigma}_{A,B\otimes C}\tilde{\alpha}_{B,C,A} = R_{A,B,C}tw_{A,B\otimes C}\epsilon^2 + \text{lower order terms,}$$

where

$$L_{A,B,C} =$$
$$a\kappa_{A,B}^2 \otimes C + b\kappa_{A,B}^2 \otimes C + c\kappa_{A,B}\kappa_{B,C} + d\kappa_{A,B}\kappa_{A,C}$$
$$+ eA \otimes \kappa_{B,C}^2 + f\kappa_{A,C}\kappa_{B,C} + g\kappa_{A,C}\kappa_{A,B}$$
$$+ h\lceil B \otimes \kappa_{A,C}^2 \rceil + a\lceil B \otimes \kappa_{A,C}^2 \rceil + \kappa_{A,B}\kappa_{A,C},$$

and

$$R_{A,B,C} =$$
$$b\kappa_{A,B}^2 \otimes C + c\kappa_{A,B}\kappa_{A,C} + d\kappa_{A,B}\kappa_{B,C} + e\lceil B \otimes \kappa_{A,C}^2 \rceil$$
$$+ f\kappa_{B,C}\kappa_{A,C} + g\kappa_{B,C}\kappa_{A,C} + hA \otimes \kappa_{B,C}^2 + a\kappa_{A,B}^2 \otimes C$$
$$+ a\kappa_{A,C}\kappa_{A,B} + a\kappa_{A,B}\kappa_{A,C} + a\kappa_{A,B}\kappa_{B,C} + a\lceil B \otimes \kappa_{A,C}^2 \rceil$$
$$+ bA \otimes \kappa_{B,C}^2 + c\kappa_{B,C}\kappa_{A,C} + d\kappa_{B,C}\kappa_{A,C} + ea\kappa_{A,B}^2 \otimes C$$
$$+ f\kappa_{A,C}\kappa_{B,C} + g\kappa_{A,C}\kappa_{A,B} + h\lceil B \otimes \kappa_{A,C}^2 \rceil.$$

In the expression for $L_{A,B,C}$ the last term is the composition of the first order terms from the two braidings.

Since the lower order terms agree, this hexagon reduces to $L_{A,B,C} = R_{A,B,C}$. Cancelling like terms, rewriting $\kappa_{A,C}\kappa_{A,B}$ using the 4T relation, and equating like terms, this reduces to

$$
\begin{aligned}
e &= 0 \\
a + c + d &= 0 \\
c + d &= 1 \\
b + h &= 0 \\
a + c + f &= 0 \\
d + g &= 0.
\end{aligned}
$$

Now, since we are in characteristic 2 and $\epsilon^4 = 0$, it follows that $\tilde{\alpha}^{-1} = \tilde{\alpha}$. Thus, by symmetry, the other hexagon reduces to the equations

$$
\begin{aligned}
e &= 0 \\
a + f + g &= 0 \\
f + g &= 1 \\
b + h &= 0 \\
a + d + g &= 0 \\
c + f &= 0.
\end{aligned}
$$

Now, from the last equation in the first set and the next to last in the second, we conclude that $a = 0$. The second and third equations of either set then yield a contradiction, since $c + d$ and $f + g$ must be both 0 and 1.

Thus we see that the second obstruction does not vanish. $\square$

In a way, this result is hardly surprising: Drinfel'd's construction of a deformed braiding and associator involved exponentiating the first order deformation of the braiding, and the resulting second order term for the deformed associator also involved a denominator of 2.

It turns out that it is better to regard this last deformation as an external deformation of $\mathcal{F}(1)$, the free symmetric compact closed category on one object generator. When we do this, we find that the situation is even more rigid than this theorem indicates: we cannot even adjoin a natural endomorphism Θ of $\otimes$ which will serve as a second order deformation term unless either additional equations are imposed on κ, or a deformation term for the associator which satisfies certain equations relating it to κ and Θ is formally adjoined. More precisely, we have:

Theorem 20.2 *If $j : \mathcal{F}(1) \to \mathcal{C}$ is the canonical inclusion of symmetric compact closed categories given by mapping, and $\mathcal{C}$ is generated as a monoidal category by the monoidal subcategory $\mathbf{VasTang}_k$ and a natural endomorphism $\Theta_{A,B}$ of $\otimes$, then the second obstruction to extending the first order deformation of $\mathcal{F}(1)$ does not vanish in $\mathcal{C}$ for any deformation with braiding of the form $[Id_{A,B} + \kappa_{A,B}\epsilon + \Theta_{A,B}]tw_{A,B}\epsilon^2$.*

sketch of proof: The proof is a very unedifying calculation: we consider a second order deformation term for the associator that is a generic linear combination of prolongations of Θ and composites of two prolongations of κ, and consider generic relations expressing $\Theta_{A\otimes B,C}$ and $\Theta_{A,B\otimes C}$ as linear combinations of prolongations of Θ for pairs of single objects and composites of two prolongations of κ, expressing Θtw as a linear combination of Θ and κ^2.

Calculating the two hexagons gives inconsistent equations on the coefficients. $\square$

On the other hand, we have:

Theorem 20.3 *Let $j : \mathcal{F}(1) \to \mathcal{C}$ be the canonical map of $\mathcal{F}(1)$ to the monoidal category $\mathcal{C}$ which is generated by the monoidal subcategory $\mathbf{VasTang}_k$, a multiprincipal natural transformation*

$$\Phi_{A,B,C} : A \otimes B \otimes C \to A \otimes B \otimes C,$$

and a natural transformation $\Theta_{A,B} : A \otimes B \to A \otimes B$, *subject to the equations*

$$\Theta_{A\otimes B,C} =$$
$$\lceil A \otimes \kappa_{B,C} B \otimes \kappa_{A,C} \rceil + A \otimes \Theta_{B,C} + \lceil B \otimes \Theta_{A,C} \rceil$$
$$+ \Phi_{A,B,C} + \lceil \Phi_{A,C,B} \rceil + \lceil \Phi_{C,A,B} \rceil$$

$$\Theta_{A,B\otimes C} =$$
$$\lceil \kappa_{A,B} \otimes C \kappa_{A,C} \otimes B \rceil + \Theta_{A,B} \otimes C + \lceil \Theta_{A,C} \otimes B \rceil$$
$$+ \Phi_{A,B,C} + \lceil \Phi_{B,A,C} \rceil + \lceil \Phi_{B,C,A} \rceil$$

$$\Theta_{I,A} = 0$$

$$\Theta_{A,I} = 0$$

$$\epsilon_A \Theta_{A,A^*} = 0$$

$$\Theta_{A^*,A}\eta_A + \Omega_{A^*,A}\eta_A + [A^* \otimes \Upsilon_A]\eta_A = 0$$

$$[A \otimes \epsilon_A][\Theta_{A,A} \otimes A^* + A \otimes \Upsilon_A][tw_{A,A} \otimes A^*][A \otimes \eta_{A^*}] = 0$$

$$[A^* \otimes \epsilon_A]\Phi_{A^*,A,A^*}[\eta_A \otimes A^*] + \Upsilon_A^* = 0$$

$$\Theta_{A,B}tw = tw\Theta_{B,A} + tw\Omega_{A,B},$$

where $\Upsilon_A = [\epsilon_A \otimes A]\Phi_{A,A^*,A}[A \otimes \eta_A]$, *and*

$$\Omega_{A,B} =$$

$$\lceil [A \otimes \epsilon_B \otimes B][\Phi_{A,B,B^*} \otimes B][A \otimes B \otimes \eta_B]\rceil$$

$$+\lceil [\epsilon_B \otimes A \otimes B][\Phi_{B,B^*,A} \otimes B]$$

$$[tw_{B \otimes B^*,A} \otimes B][A \otimes B \otimes \eta_B]\rceil$$

$$+\lceil [A \otimes \epsilon_B \otimes B][tw_{A,B} \otimes B^* \otimes B]$$

$$[\Phi_{B,A,B^*} \otimes B][tw_{B,A} \otimes \eta_B]\rceil$$

$$+\lceil [A \otimes \epsilon_{A^*} \otimes B][A \otimes \Phi_{A^*,A,B}][\eta_{A^*} \otimes A \otimes B]\rceil$$

$$+\lceil [A \otimes B \otimes \epsilon_{A^*}][A \otimes \Phi_{B,A^*,A}]$$

$$[A \otimes tw_{B,A^* \otimes A}][\eta_{A^*} \otimes A \otimes B]\rceil$$

$$+\lceil [A \otimes B \otimes \epsilon_{A^*}][A \otimes tw_{B,A^*} \otimes B]$$

$$[A \otimes \Phi_{A^*,B,A}][\eta_{A^*} \otimes twA, B]\rceil.$$

Then

$$\tilde{\sigma} = [Id + \kappa\epsilon + \Theta\epsilon^2]tw$$

$$\tilde{\alpha} = Id + \Phi\epsilon^2$$

defines a second order extrinsic deformation of $\mathcal{F}(1)$.

In this theorem and in the following the $\lceil\ \rceil$ operator includes the coherence maps for the *symmetric* monoidal structure, so that inputs can be permuted to match sources and targets. (Observe that all natural transformations are paracoherent, so this introduces no ambiguities.)

proof: The proof consists of two parts: first verifying the hexagons, pentagon and triangles; and second checking that the equations given do not collapse the image of $\mathcal{F}(1)$.

In the first part it suffices to check the terms of degree 2 in ϵ.

The pentagon follows immediately from the condition that Φ be multiprincipal. The hexagons and triangles can be readily verified by direct calculation (indeed computing, the degree 2 terms of the hexagons taking the last relation as an *Ansatz* yields the first two relations, while computing the triangles yields the third and fourth.). The fifth through seventh relations are imposed to give unframed knot invariants – and correspond, more or less, to the possible instances of a Reidemeister move of type $\Omega.1$. The theorem will hold without them, though a few details of the following will change. Indeed, these too can be read off the categorical versions of $\Omega.1$ in the presence of the last relation. The eighth relation is imposed so that both duality conditions will hold.

The second part of the proof consists of observing that if we define the degree of a map in $\mathcal{C}$ so that the degree of maps in $\mathcal{F}(1)$ is 0, the degree of an instance of κ is 1, the degree of an instance of Θ or Φ is 2, and degree is additive with respect to both composition and $\otimes$, then degree is well defined, and all relations imposed by the structure of **VasTang**$_k$ and the statement of the theorem are on maps of degree 1 or 2. Thus no new relations are imposed on the image of $\mathcal{F}(1)$. $\square$

Observe that the eighth relation and the presence of Υ_A in the other relations arise from the need to replace η_A with

$$\tilde{\eta}_A = \eta_A + [A^* \otimes \Upsilon_A]\eta_A \epsilon^2$$

to preserve the duality. (The analogous relation to the eighth relation in Drinfel'd's characteristic 0 construction holds as a result of the symmetry properties of the second order term in the associator.)

Chapter 21

Categorical Deformations as Proper Generalizations of Classical Notions

If, by abuse of notation, we denote the lax monoidal functor defining (or defined by) A as $A : ! \to R\text{-}\mathbf{mod}$, we then have

Theorem 21.1 $X^\bullet(A)$ *is isomorphic to* $C^\bullet(A, A)$ *as differential graded R-algebras with respect to* $\cup$, *and, moreover, the isomorphism is an isomorphism of pre-Lie structures. That is, it preserves the indexed products* $\langle -, - \rangle_i$ *of the pre-Lie structure.*

proof: Now, $X^n(A) = Nat(^n\otimes(A^n), A(\otimes^n))$, but since the source category has only one object and one map, this reduces to

$$Hom_{R-mod}(^\otimes(A^n), A) = C^n(A, A).$$

For the rest of the preservation properties, it suffices to observe that the formula for the structure on $X^\bullet(F)$ specializes to an element-free description of the corresponding structure on the Hochschild complex.

For the coboundary we have

209

$$\delta(\phi)_{!_0,\ldots,!_n} = \lceil A \otimes \phi_{!_1,\ldots,!_n} \rceil + \sum_{i=1}^{n} (-1)^i \lceil \phi_{!_0,\ldots,!_{i-1} \otimes !_i,\ldots,!_n} \rceil$$
$$+(-1)^{n+1} \lceil \phi_{!_0,\ldots,!_{n-1}} \otimes A \rceil,$$

where the subscripts on the !'s merely distinguish different copies. We must now consider the effects of the operator $\lceil \rceil$ on each term. Since each term consists of $\lceil f \rceil$ for a single map f, $\lceil \rceil$ represents pre- and post-composition by coherence maps for $R - mod$ or the lax monoidal functor A or images under A of coherence maps for !. This last is trivial since the only map in ! is an identity map. For the first (resp. last) terms, it is easy to see that the required composite is precomposition by the identity map and postcomposition by $\tilde{A} = m$, the multiplication on A (resp. precomposition by the complete right reassociation coherence map in R-**mod** and postcomposition by $\tilde{A} = m$), and thus these terms equal the corresponding terms of the Hochschild coboundary. For all other terms, there is no postcomposition, only precomposition by the appropriate coherence map in R-**mod** followed by a prolongation of $\tilde{A}$, with $\tilde{A}$ applied to the $i - 1^{st}$ and i^{th} tensorands. A moment's reflection shows that these are precisely the other terms of the Hochschild coboundary.

Similar considerations show that the cup product and each of the pre-Lie products are also preserved. $\square$

A similar result holds for oplax monoidal functors and counital coassociative coalgebras.

Thus, we see that our theory is a proper generalization of the deformation theory of Gerstenhaber for rings and algebras [23], in the sense that the classical theory is included as a special case.

More than this, with the present theory, we are able to give a cohomological account of deformations of a unital associative algebra (resp. counital coassociative coalgebra) that give rise to "quasi-associative" algebras (resp. "quasi-coassociative" coalgebras): we can consider the total deformation complex of the lax (resp. oplax) monoidal functor. If one wishes to consider the most general possible "quasi-algebras" one must consider deformations of a monoidal subcategory of the ambient category of modules which contains the algebra, its multiplication, and its unit map. Plainly, cutting down to only the objects generated monoidally by the algebra itself allows for more deformations. On the other hand, when considering which maps lie in the subcategory, there is a trade-off between losing maps which might serve as deformation terms and losing restrictions imposed by the naturality condition.

The theory also includes group cohomology with trivial coefficients as a special case. Consider the following example:

Fix a group G and a field K. Let $\mathcal{C}_{G,K}$ be the category of G graded K-vectorspaces. Objects are of the form $\oplus_{g \in G} M_g$, where M_g is the K-vectorspace of degree g elements, and the dimension of the direct sum is finite, with degree preserving K-linear maps as maps. The category is equipped with a monoidal product

$$[\oplus_{g \in G} M_g] \otimes [\oplus_{h \in G} N_h] = \oplus_{k \in G} [\oplus_{gh=k} M_g \otimes_K N_g]$$

with unit K_e, and structural natural transformations induced by the structural transformations for $\otimes$ and $\oplus$ on $K - v.s.$.

Observe that $\mathcal{C}_{G,K}$ is equipped with a monoidal forgetful functor U to $K - v.s.$

Now, $\mathcal{C}_{G,K}$ is easily seen to be semisimple with generating objects $K_g \; g \in G$. Thus, it is easy to see, using the properties of finite direct sums (as iterated biproducts, cf. Mac Lane [40]) that to specify a natural transformation between two functors from $\mathcal{C}_{G,K}^n$ to $\mathcal{C}_{G,K}$ or $K - v.s.$, it suffices to specify the components at each object of the form $[K_{g_1}, \ldots, K_{g_n}]$. Moreover, in the case of functors both of which are composites of prolonged monoidal products and forgetful functors, the image of any such object will be a simple object (isomorphic to either $K_{g_1 \ldots g_n}$ in $\mathcal{C}_{G,K}$ or K in $K - v.s.$). Thus, to specify the component of the natural transformation is simply to specify a scalar.

By identifying $[K_{g_1}, \ldots, K_{g_n}]$ with $[g_1, \ldots, g_n])$, we see that

$$X^n(\mathcal{C}_{G,K}) \cong C^n(G, K) \cong X^n(U).$$

It is easy to verify that this is, in fact, a chain map. Thus, we have as a corollary to our main results:

Corollary 21.2 *The semigroupal deformations of $\mathcal{C}_{G,K}$ are classified by $H^3(G,K)$.*

and

Corollary 21.3 *The deformations of the strong monoidal functor*

$$U : \mathcal{C}_{G,K} \to K - v.s.$$

are classified by $H^2(G,K)$.

Chapter 22

Open Questions

22.1　Functorial Knot Theory

The chief problem surrounding functorial knot theory is to give a general and satisfactory relationship between the rich families of knot and link invariants it provides and the more geometric problems of knot theory as classically studied. Some success has been had in scattered instances (for example, Kirby and Melvin [36]), while some functorial knot invariants have been deliberately constructed using homotopy theoretic constructions so that an immediate homotopic interpretation is given along with the invariant (cf. Yetter [62]). On the whole, though, the data about knots and links encoded by functorial knot invariants remains tantalizingly ungeometric.

A second problem that has remained open since the discovery of the HOMFLY and Kauffman polynomials is to give a satisfactory explanation for the existence of the values not corresponding to the representations of quantum groups. For example, in the case of the HOMFLY polyomial, the special values associated to the fundamental representations of $U_q(sl_n)$, and the fact that skein theory in the manner of Kauffman and Lins [31] can be used to reconstruct (for generic q) the complete categories of finite dimensional representations of $U_q(sl_n)$ suggest that if we write the HOMFLY polynomial with the normalization

213

$x = q^\zeta$, $z = q^{\frac{1}{2}} - q^{-\frac{1}{2}}$, for a complex parameter ζ, we might regard the entire invariant as arising from the "fundamental representation of $U_q(sl_\zeta)$". The difficulty is to make sense of this.

There is precedent for making the leap from a parameter which constitutes a dimension to a complex parameter: the "dimensional regularization" schemes used by physicists to remove the divergences prevalent in the Feynmanological approach to QFT.

22.2   Deformation Theory

There are perhaps too many open questions regarding the deformation theory of monoidal categories and monoidal functors. This is hardly surprising given on the one hand its recent advent, and on the other the fact that the theory is at least as rich as Gerstenhaber's deformation theory for associative algebras.

The questions which occur to the author in this regard may be classified into three types: those dealing with the theory itself, those seeking clarification of the relationship of the theory to the deformation theories of Gerstenhaber [23, 24] and of Gerstenhaber and Schack [25], and those dealing with the theory's relationship to Vassiliev theory.

Of questions about the theory itself, two seem most important: the question of functoriality properties for the deformation complexes, and the problem of effective computation.

No naive functoriality properties for the deformation complexes (even considered as objects in a derived category) present themselves beyond those of Theorems 13.6 and 13.7. Rather, to any monoidal functor, one has a cospan of cochain complexes formed by the maps introduced in Chapter 14, and a distinguished triangle formed by the direct sum of the deformation complexes for the source and target, the deformation complex for the functor, and the cone (whose third cohomology classifies total deformations). This distinguished triangle, together with the projections, gives rise to a span (of maps of degree $+1$). It is unclear whether the composition of monoidal functors carries over to the composition of cospans (via pushouts), to composition of spans (via pullback), to both, or to neither. One might hope that one of these conditions holds on the nose, but it might be necessary to pass to the derived category.

It is not even clear whether there are simple functoriality properties for any class of monoidal functors more general than monoidal equivalences.

The lack of naive functoriality is not surprising. The deformation complexes of algebras lack good functoriality properties with respect to maps of algebras. In the algebra case, one can pass to Hochschild cohomology with coefficients in a module to obtain a construction with good functoriality properties in each of two variables. An analogous construction has been developed by the author in unpublished work [58], but at present is not well-understood.

The second problem is that of effective calculation. It would be highly desirable to have a construction which, given a monoidal category or monoidal functor, provides a simplicial complex whose simplicial cochain complex is isomorphic, or at least chain homotopic or quasi-isomorphic to the deformation complex. This, however, would not be the complete solution to computation, since it would not determine the pre-Lie structure needed to compute obstructions.

The need to model the pre-Lie structure suggests another approach: find a construction which, given a monoidal category or functor, constructs an associative algebra whose Hochschild complex is isomorphic to the deformation complex of the category or functor by an isomorphism which preserves the pre-Lie structure.

Finding such a construction is only the first open question regarding the relationship between categorical deformation theory and classical algebraic deformation theory. It would also be highly desirable to understand the relationship between the deformations of a bialgebra in the sense of Gerstenhaber and Schack [25] and the deformations of its category of modules, or of the underlying functor $U : R\text{-}\mathbf{mod} \to K\text{-}\mathbf{v.s.}$. Likewise, the same relationship should be investigated in the dual setting for categories of comodules. This latter is almost more important since reconstruction theorems allow any K-linear monoidal category fibered over K-v.s. (and satisfying certain exactness properties) to be realized as a category of comodules over a bialgebra.

A third approach would be to construct in the context of monoidal categories and monoidal functors analogues of the standard machinery of homological algebra: resolutions in terms of some nice class of acyclic objects, ideally projectives in some suitable category. Some progress on this approach has been made by the author in unpublished work [58], in which it is shown that under suitable exactness hypotheses, including the abelianness of the target category, analogues of modules for lax monoidal functors more general than those corresponding to algebras form an abelian category. The deformation cohomology defined herein can then be extended to a notion of the cohomology of a monoidal functor F with coefficients in an F-bimodule, while the apparent problem of the cochain groups being defined in terms of functors from various Deligne-powers of the source category can be removed by the use of left Kan-extensions (cf. [40, 58]).

Finally, the intimate relationship between categorical deformation theory and Vassiliev theory suggests a number of possible lines of development. The realization of the Kontsevich integral as a deformation in the category **VasTang$_R$** has already provided a satisfactory explanation for the need to adjust the direct formulation to obtain isotopy invariance — the need to preserve duality once the associator has been deformed — and has suggested the correct way to make the adjustment in light of the result of Vogel [56] that semi-simple categories will not suffice to generate all Vassiliev invariants. One program which should be undertaken is the use of cohomological and deformation theoretic techniques to study Vassiliev invariants, and particularly the question of integrability of weight systems in a systematic way. A second program, already begun in a small in Chapter 20, is to use deformation theory to undertake the study of Vassiliev invariants valued in finite fields, where the transcendental methods of Drinfel'd and Kontsevich cannot be applied.

Only time will tell whether the specific issues and questions raised in this monograph will be of any enduring importance. What cannot be doubted is that the direct conversion of geometric data into algebra, as exemplified by functorial knot theory, has established the theory of categories with structure as an important branch of algebra and opened new mathematical vistas which will occupy topologists, algebraists, and others for many years to come.

Bibliography

[1] J. W. Alexander. A lemma on systems of knotted curves. *Proc. Nat. Acad. Sci. U.S.A.*, 9:93–95, 1923.

[2] E. Artin. Theory of braids. *Ann. of Math.*, 48:101–126, 1947.

[3] M. Atiyah. Topological quantum field theories. *Publ. I.H.E.S.*, 68:175–186, 1988.

[4] M.F. Atiyah and I.G. Macdonald. *Introduction to Commutative Algebra*. Addison-Wesley, Reading, MA, 1969.

[5] Dror Bar-Natan. Non-associative tangles. In W. H. Kasez, editor, *Geometric Topology*, pages 139–183. American Mathematical Society. (proceedings of the Georgia International Topology Conference).

[6] Dror Bar-Natan. On the Vassiliev knot invariants. *Topology*, 34(2):423–472, 1995.

[7] Dror Bar-Natan. On associators and the Grothendieck-Teichmuller group, I. *Selecta Mathematica, New Series*, 4:183–212, 1998.

[8] J. Birman and X.-S. Lin. Knot polynomials and Vassiliev's invariant. *Invent. Math.*, 111, 1993.

[9] Joan Birman. *Braids, Links, and Mapping Class Groups*. Princeton Univ. Press, Princeton, NJ, 1975.

[10] R. D. Brandt, W. B. R. Lickorish, and K. C. Millet. A polynomial invariant for unoriented knots and links. *Invent. Math.*, 84:563–573, 1986.

[11] Gerhard Burde and Heiner Zieschang. *Knots.* Walter de Gruyter, Berlin, 1985.

[12] L. Crane and I. B. Frenkel. Four-dimensional topological quantum field theory, Hopf categories, and the canonical bases. *J. Math. Phys.*, 35(10):5136–5154, 1994.

[13] L. Crane and D.N. Yetter. Deformations of (bi)tensor categories. *Cahier de Topologie et Géometrie Differentielle Catégorique*, 1998.

[14] A.A. Davydov. Twisting of monoidal structures. preprint, 1997.

[15] P. Deligne. letter to D.N. Yetter dated 20 January 1990.

[16] V.G. Drinfel'd. Hopf algebras and the quantum Yang-Baxter equation. *Dokl. Akad. Nauk SSSR*, 283(5):1060–1064, 1985. (Russian).

[17] V.G. Drinfel'd. Quantum groups. In A.M. Gleason, editor, *Proceedings of the International Congress of Mathematicians, Berkeley, 1986*, pages 789–820. American Mathematical Society, 1987.

[18] V.G. Drinfel'd. Quasi-hopf algebras. *Leningrad Math. J.*, 1:1419–1457, 1990.

[19] V.G. Drinfel'd. On quasitriangular quasi-hopf algebras and a group closely connected with $Gal(\overline{\mathbf{Q}}/\mathbf{Q})$. *Leningrad Math. J.*, 2:829–860, 1991.

[20] B. Eckmann and P.J. Hilton. Group-like structures in general categories, I. Multiplications and comultiplications. *Math. Ann.*, 145:227–255, 1962.

[21] D.B.A. Epstein. Functors between tensored categories. *Invent. Math.*, 1:221–228, 1966.

[22] P.J. Freyd and D.N. Yetter. Coherence theorems via knot theory. *Journal of Pure and Applied Algebra*, 78:49–76, 1992.

[23] M. Gerstenhaber. The cohomology of an associative ring. *Ann. of Math.*, 78(2):267–288, 1963.

[24] M. Gerstenhaber. On the deformation of rings and algebras. *Ann. of Math.*, 79(1):59–103, 1964.

[25] M. Gerstenhaber and D. Schack. Algebraic cohomology and deformation theory. In M. Hazewinkel and M. Gerstenhaber, editors, *Deformation Theory of Algebras and Structure and Applications*, pages 11–264. Kluwer, 1988.

[26] V. Goryunov. Vassiliev invariants of knots in $\mathbf{R}^3$ and in a solid torus. Preprint, 1995.

[27] Guilleman and Pollack. *Differential Topology*. Prentice Hall, New York, 1979.

[28] M. Jimbo. A q-difference analogue of $U(\mathfrak{g})$ and the Yang-Baxter equation. *Lett. Math. Phys.*, 10(1):63–69, 1985.

[29] V.F.R. Jones. A polynomial invariant of knots via vonNeumann algebras. *Bull. A.M.S.*, 12:103–111, 1985.

[30] A. Joyal and R. Street. Braided tensor categories. *Adv. in Math.*, 102:20–78, 1993.

[31] L. H. Kauffman and S. L. Lins. *Temperley-Lieb recoupling theory and invariants of 3-manifolds*. Princeton Univ. Press, Princeton, 1994. Annals of Mathematic Studies, no. 134.

[32] L.H. Kauffman. An invariant of regular isotopy. *Trans. A.M.S.*, 318(2):317–371, 1990.

[33] G.M. Kelly. *Basic Concepts of Enriched Category Theory.* Cambridge University Press, Cambridge, 1982. LMS Lecture Notes.

[34] G.M. Kelly and Laplaza. Coherence for compact closed categories. *Journal of Pure and Applied Algebra*, 19:193–213, 1980.

[35] Robion Kirby. A calculus for framed links in S^3. *Invent. Math.*, 45:36–56, 1978.

[36] Robion Kirby and Paul Melvin. The 3-manifold invariants of Witten and Reshetikhin-Turaev for $sl(2, \mathbb{C})$. *Invent. Math.*, 105:473–545, 1991.

[37] M. Kontsevich. Vassiliev's knot invariant. *Adv. Soviet Math.*, 16(2):137–150, 1993. I. M. Gelfand Seminar.

[38] V. Lyubashenko. The triangulated Hopf category $n_+SL(2)$, 1999/2000. (e-print) abstract at http://xxx.lanl.gov/abs/math.QA/9904108.

[39] Saunders Mac Lane. Natural associativity and commutativity. *Rice Univ. Studies*, 49:28–46, 1963.

[40] Saunders Mac Lane. *Categories for the Working Mathematician, second edition.* Springer-Verlag, Berlin, 1998.

[41] A. A. Markov. Uber die frei Aquivalenz geschlossen Zöpfe. *Mat. Sb.*, 1:73–78, 1935.

[42] P.J. Freyd, D.N. Yetter; W.B.R. Lickorish, K.C. Millet; J. Hoste; and A. Ocneanu. A new polynomial invariant of knots and links. *Bull. A.M.S.*, 12:239–246, 1985.

[43] J. Przytcki and Traczyk P. Conway algebras and skein equivalence of links. *Proc. Amer. Math. Soc.*, 100:744–748, 1987.

[44] Kurt Reidemeister. *Knottentheorie.* Springer-Verlag, Berlin, 1932.

[45] N. Yu. Reshetikhin. Quantized universal enveloping algebras, the Yang-Baxter equation and invariants of links, I. *LOMI Preprints*, E-4-87, 1988.

[46] N. Yu. Reshetikhin and V. G. Turaev. Ribbon graphs and their invariants derived from quantum groups. *Comm. Math. Phys.*, 127:1–26, 1990.

[47] N. Saavedra Rivano. *Catégories Tannakiennes*. Springer-Verlag, Berlin, 1972. Lecture Notes in Mathematics, vol. 265.

[48] M.-C. Shum. *Tortile Tensor Categories*. PhD thesis, Macquarie University, 1989.

[49] M.-C. Shum. Tortile tensor categories. *Journal of Pure and Applied Algebra*, 93:57–110, 1994.

[50] Michael Spivak. *Differential Geometry, a Comprehensive Introduction, vol. 1*. Publish or Perish, Houston,TX, 1973.

[51] T. Stanford. Finite-type invariants of knots, links, and graphs. *Topology*, 35(4):1027–1050, 1996.

[52] J. Stasheff. Homotopy associativity of H-spaces, I, II. *Trans. A.M.S.*, 108:275–292, 293–312, 1963.

[53] V. Turaev. The category of oriented tangles and its representations. *Funct. Anal. Appl.*, 23:254–255, 1989.

[54] K.-H. Ulbrich. On Hopf algebras and rigid monoidal categories. *Israel J. Math.*, 72:252–256, 1990.

[55] V.A. Vassiliev. Cohomology of knot spaces. In V.I. Arnold, editor, *Theory of Singularities and its Applications*, pages 23–69. American Mathematical Society, Providence, RI, 1990. A.M.S. Adv. in Sov. Math. vol. 1.

[56] P. Vogel. Algebraic structures of modules of diagrams. *Inventiones Mathematica*. (to appear).

[57] C.A. Weibel. *An introduction to homological algebra.* Cambridge Univ. Press, Cambridge, 1994.

[58] D.N. Yetter. Abelian categories of modules over a (lax) monoidal functor. (in preparation).

[59] D.N. Yetter. Markov algebra. In J.S. Birman and A. Libgober, editors, *Braids, Santa Cruz, 1986*, pages 705–730. American Mathematical Society, 1988. Contemporary Mathematics, vol. 78.

[60] D.N. Yetter. Quantum groups and representations of monoidal categories. *Math. Proc. Camb. Phil. Soc.*, 108:261–290, 1990.

[61] D.N. Yetter. Framed tangles and a theorem of Deligne on braided deformations of Tannakian categories. In M. Gerstenhaber and J. Stasheff, editors, *Deformation Theory and Quantum Groups with Applications to Mathematical Physics*, pages 325–350. American Mathematical Society, 1992. A.M.S. Contemp. Math. vol. 134.

[62] D.N. Yetter. Topological quantum field theories associated to finite groups and crossed G-sets. *Journal of Knot Theory and its Ramifications*, 1(1):1–20, 1992.

[63] D.N. Yetter. Braided deformations of monoidal categories and Vassiliev invariants. In E. Getzler and M. Kapranov, editors, *Higher Category Theory*, pages 117–134. American Mathematical Society, 1998. A.M.S. Contemp. Math. vol. 230.

Index